KV-062-789

Ageing Pipelines

IMechE Conference Transactions

Ageing Pipelines

Optimizing the Management and Operation: Low Pressure – High Pressure

11–13 October 1999
Newcastle Civic Centre, UK

Organized by
The Pressure Systems Group of the
Institution of Mechanical Engineers (IMechE)

IMechE Conference Transactions 1999–8

Published by Professional Engineering Publishing Limited for The Institution of Mechanical Engineers, Bury St Edmunds and London, UK.

First Published 1999

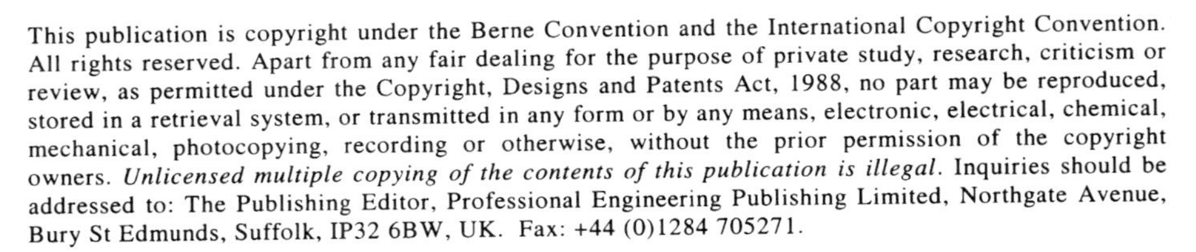

ISSN 1356–1448
ISBN 1 86058 232 X

A CIP catalogue record for this book is available from the British Library.

Printed by The Book Company, Ipswich, Suffolk, UK.

Conference Organizing Committee

M Brown (Chairman)
Transco

J Darlaston
Strutech Consultancy

D Dickson
Advanced Engineering Solutions Limited

P Hopkins
Andrew Palmer and Associates

C Reed
Northumbrian Water

N Townsend
Northern Energy Initiative

Sponsored by

United Kingdom On-shore Pipeline Operation Association (UKOPA)
BG Technology
Transco
BGE (UK) Limited
Newcastle City Council

Co-sponsored by

Pipeline Industries Guild
Institution of Corrosion
Institute of Materials
Chartered Institution of Water and Environment Management
Association of British Offshore Industries (ABOI)
Institute of Petroleum
Northern Energy Initiative

Related Titles of Interest

Title	Editor/Author	ISBN
Engineers' Data Book	Clifford Matthews	1 86058 175 7
Engineering System Safety	G J Terry	0 85298 781 1
The Economic Management of Physical Assets	N W Hodges	0 85298 958 X
An Engineer's Guide to Pipe Joints	G Thompson	1 86058 081 5
Practical Guide to Engineering Failure Investigation	Clifford Matthews	1 86058 086 6
Assuring its Safe: Integrating Structural Integrity, Inspection, and Monitoring into Safety and Risk Assessment	IMechE Conference	1 86058 147 1
Remanent Life Prediction	IMechE Seminar	1 86058 154 4
Water Pipeline Systems (BHR Group Publication 23)	R Chilton	1 86058 088 2

For the full range of titles published by Professional Engineering Publishing contact:

Sales Department
Professional Engineering Publishing Limited
Northgate Avenue
Bury St Edmunds
Suffolk
IP32 6BW
UK

Tel: +44 (0)1284 724384
Fax: +44 (0)1284 718692

Contents

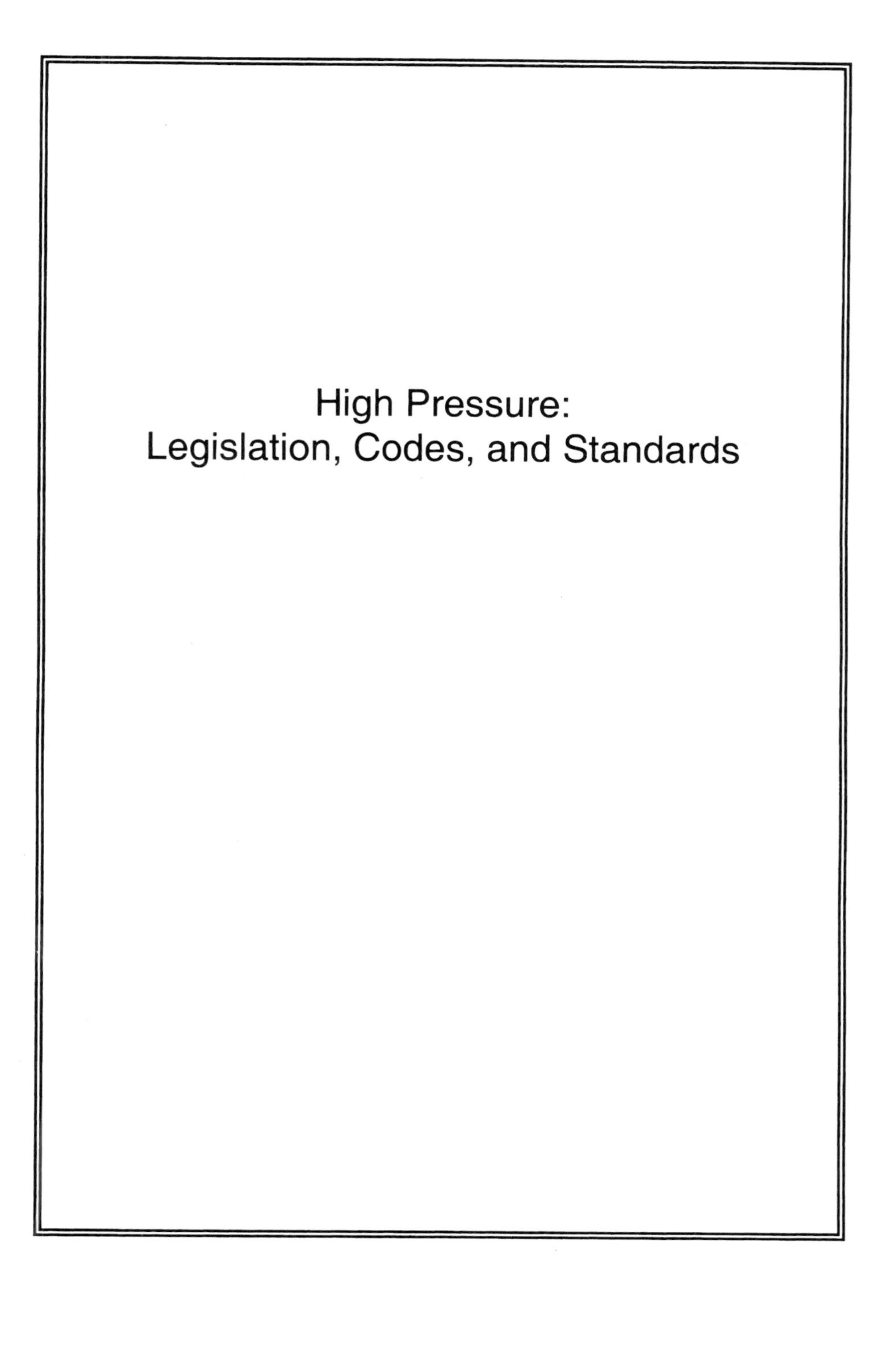

High Pressure:
Legislation, Codes, and Standards

C571/022/99

IGE/TD/1: Steel pipelines for high-pressure gas transmission – the need for change

G SENIOR
Transco, Tyne and Wear, UK
R N KNOTT
Consultant, Derby, UK

ABSTRACT

Any recognised pipeline standard must ensure that it keeps up to date with good practice and developing technologies. Consequently, the Institution of Gas Engineers (IGE) in the UK periodically reviews its various recommendations, as they are called, and issues new editions when appropriate. The current IGE "recommendations" for gas transmission pipelines are detailed in IGE/TD/1[1]. This document, used to design and operate pipelines in the UK, is currently under review by the IGE and is the subject of this paper.

As this conference is about ageing pipelines, this paper concentrates on some of the challenges faced by today's pipeline operator when managing a pipeline through its entire life cycle. These challenges include changes in duty, integrity, operation / maintenance and legislation.

This paper explains how these are being considered in the proposed new edition of IGE/TD/1.

1. IGE/TD/1 PANEL

The IGE has established several technical panels, which undertake work in a particular field or subject. The Panel PTD/1 is responsible for steel pipelines carrying gas at high pressures, i.e. above 7 bar. It reports to the Transmission and Distribution Committee, TDC, and its principal function is the development of the Institution's recommendations for "Steel pipelines for high pressure gas transmission", or more commonly called TD/1.

1.1 History

The panel has been in existence for many years under a number of different chairmen. Over the years nearly a hundred engineers, all involved in gas transmission, have been members.

In 1967 the IGE published its first recommendations relating to steel pipelines as publication 674. 1970 saw the first version of IGE/TD/1. An additional section covering pipeline sleeving was produced in 1976, as publication 674B. The various sections of 674 A B C D were combined and republished as Edition 1 of IGE/TD/1 in 1977. Later the same year the first section (Section 4: Design) of the Edition 2 was produced. The complete Edition 2 was finally printed in 1984. In 1993, a complete revision was undertaken and Edition 3(1) was published.

After the IGE had published Edition 3 in 1993, the panel became dormant, as there were no major matters of interest. However, in 1996 it was reconvened to consider extending the recommendations to cover 1219mm (48 inch) pipelines. At the same time, British Gas was undergoing a major reorganisation and the majority of the panel members either left the industry or took up alternative positions. The reconvened panel required membership to represent the UK Gas Industry. The panel is currently chaired by Robin Knott, an independent consultant, and consists of independent representation from Penspen Ltd, Kvaerner John Brown Ltd, Transco, the HSE, British Steel, Alfred McAlpine Services, BG Technology, and Pipeline Integrity International Ltd., with the Secretary from the IGE secretariat.

1.2 Current Status

Since the publication of Edition 3, the overall situation has moved on: (i) Pipeline design technology, legislation and the requirements of operators have changed; (ii) CEN(2) and ISO(3) are producing new pipeline standards; (iii) Risk analysis has become more formalised, see IGE/SR/24(4) and new techniques such as "Limit State" design are developing(5, 6, 7), and (iv) European standards(2) now considers the break between Distribution and Transmission to be 16 bar instead of the UK standard of 7 bar.

Since 1996, the panel has considered a number of different topics for possible inclusion into TD/1; many because users have identified specific needs. Two such topics are the extension of TD/1 to include pipeline diameters of up to 1219 mm, and the inclusion of "The handling of steel pipes and fittings", which was a separate document (IGE/TD/6), into TD/1. These two changes have recently been issued as separate supplements(8, 9).

During this process, the panel identified several topics where the latest technology indicates that benefits will be accrued if these were adopted into a pipeline specification. It is these topics which form the basis of this paper with respect to ageing pipelines.

At the time of writing this paper there are four main work groups which are considering new issues for TD/1. These groups are covering: design, materials, operation and maintenance, and inspection/commissioning. A brief summary is now provided of the issues that are currently being progressed. The work groups are still very active at present (July 1999), therefore an outline of work completed to date, or to be completed, is given below, with a further verbal progress update at the time of the conference in October 1999.

The intention at present is for the updated document to be released as IGE/TD/1 Edition 4 in October 2000[1].

---

[1] It is worth noting that the IGE is also in the process of redrafting TD/3(10), TD/9(11) and TD/10(12) (to be

2. WHAT ARE THE CONSIDERATIONS FOR AN AGEING PIPELINE SYSTEM ?

As pipelines become older, the pipeline operator has several new problems to consider, such as:

- design no longer valid, e.g. increasing the Maximum Pipeline Operating Pressure (MPOP)[7, 14] to satisfy increased throughput or storage requirements, or the pipeline being subject to additional stresses above those induced by internal pressure,
- change in infrastructure from the as-built condition, e.g. new developments in the vicinity of the pipeline which could result in population or building proximity infringements,
- change in integrity, e.g. time dependent modes such as corrosion and fatigue, or random modes such as mechanical damage,
- change in operation and maintenance, e.g. as operational experience is gained then a natural progression is to review inspection / surveillance frequencies to determine optimum levels of frequency,
- modifications to systems.

The approach being taken by the TD/1 Panel to address such issues is now reviewed.

2.1 Uprating Pipelines

TD/1 already provides guidance on the technical issues that need to be considered when uprating[2] an existing transmission pipelines, but this guidance, at present, is restricted to a maximum design factor of 0.72[3].

However, a supplement is currently being drafted by the IGE to permit uprating of transmission pipelines to a design factor of 0.8. This supplement will be at the forefront of current pipeline standards because it will detail a methodology for uprating pipelines to a design factor of 0.8. The outline of this approach is given in figure 1.

In simple terms, the supplement requires that appropriate assessments should be performed to demonstrate that the increase in failure probability of the pipeline as a result of the uprating is acceptably small. This requires the following:

- identification of all credible failure modes, and
- an assessment of the proportionate increase in failure probability on uprating, for each failure mechanism

---

combined into a new document, TD/13) and TD/12[13], with completion of all documents aimed for October 2000.

[2] Uprating is defined in TD/1 as "Increasing the maximum operating stress in a pipeline above that currently regarded as the maximum permissible, after taking whatever measures are necessary to ensure, so far as is reasonably practicable, safe operation at the higher stress".

[3] Calculation of the design factor in TD/1 (Section 6.4.2) utilizes **minimum** wall thickness.

In addition to the above, there is also a requirement to perform the more standard assessments within TD/1, as illustrated in Figure 1.

2.2 Additional Stresses

The current edition of IGE/TD/1 requires that ground movement/additional stresses should be taken into account both when designing the pipeline and also assessing stresses while in service. The code does not however prescribe how these assessments should be performed. It is the intention to detail an approach that allows a comparison to be made between actual combined stresses against recognised criteria. This approach is to build on the Dutch code, NEN 3650(15), which defines a limit state approach in assessing the acceptability of combined stresses.

A further paper(16) at this conference expands on this area by considering the influence which soil loading and restraint has on pipeline stresses.

2.3 Changes in Integrity

Two major concerns to the pipeline operator, in terms of time dependent failure mechanisms, are corrosion and fatigue.

2.3.1 Corrosion

TD/1 currently addresses corrosion by recommending the use of high-resolution internal inspection techniques at a frequency of 10 years, combined with surface inspection techniques. However, parts of the pipeline industry already embrace risk-based techniques, where due allowance is given to the following:

- the experience of one operator is that internal inspection detects mainly external corrosion defects and pre-commissioning defects in onshore gas transmission pipelines,
- the most likely failure mode for corrosion in onshore natural gas pipelines is leakage, not rupture,
- gas leaks from corrosion are unlikely to represent a significant risk to the public, and
- pre-commissioning defects, having survived an initial high level[4] hydrotest, present no significant hazard

It is then possible to develop a risk-based approach to optimise inspection frequencies by taking the above into account. At the time of writing this paper, a background paper is being written by one UK operator to explain how they have developed such an approach.

2.3.2 Fatigue

Guidance is already provided in TD/1 on the determination of acceptable levels of pressure cycling before revalidation is required. This work was based on an assumption that a defect which could survive the high level hydrotest actually existed, and that the geometry of this defect was infinite in length. Recent work by BG Technology indicates that a theoretical surviving defect, which is shorter and deeper than the above, would actually have a shorter

---

[4] In this context, high level hydrotest means to a stress level of at least 90% SMYS.

fatigue life than the infinite long defect. This has resulted in a recent review of the fatigue guidance given in TD/1, which will be reported at the conference.

2.4 Change in Infrastructure

During the lifetime of a pipeline it is likely that encroachments will occur resulting in building infringements or population density infringements. TD/1 Edition 3 addresses this by stating that a pipeline audit is performed every 4 years to identify any such infringements. The use of risk assessment is then permitted to determine if these infringements are acceptable, even though they may breech deterministic rules within the code. The clauses of the standard which refer to the use of risk assessment are in need of review to achieve clarity in the following areas:

- use of individual risk as the most appropriate measure of pipeline risk,
- consideration to the use of societal risk,
- include methodology on risk assessment, referring to IGE/SR/24, and
- consideration of the ALARP[5] principle

Details of the above will, once again, be presented at the conference.

A further area under review is that of the difference between S[6] and T[6] areas. TD/1 defines the interface between R[6] and S areas, based on the ASME B31.8 definition, as 1 person per acre, which translates to 2.5 persons per hectare. It is fair to say that this is very much a "rule of thumb". The implications of having a clear definition of a T area should not be underestimated, as any pipeline which changes from an S to a T area could, under current recommendations, require downrating to 7 bar. Work is currently underway to attempt to clarify this boundary; once again this work will be updated for the conference.

2.5 1219mm Diameter Pipelines

Supplement 2[(9)] of TD/1 details the additional requirements to extend the document scope to cover 1219mm pipelines. Specifically, the supplement provides further details on the following:

- least nominal wall thickness,
- new proximity distance[7] for pipeline diameters up to 1219mm,
- maximum design pressure for S area pipelines, based on individual risk levels, and
- pressure corrections for temperature variations during pressure testing.

---

[5] The principle of ALARP, as low as reasonably practicable, is defined in [(4)] as " ... all reasonably practicable measures will be taken in respect of risks which lie in the "tolerable" zone to reduce them further until the cost of further risk reduction is grossly disproportionate to the benefit".

[6] Type R - "Rural areas with a population density not exceeding 2.5 persons per hectare".
Type S - "Areas intermediate in character between Types R and T in which the population density exceeds 2.5 persons per hectare and which may be extensively developed with residential properties, schools, shops, etc".
Type T - "Central areas of towns or cities, with a high population density, many multi-storey buildings, dense traffic and numerous underground services".

[7] Minimum distance permissible between the pipeline and any occupied building or traffic route.

3. COMMENTS

The IGE will welcome any suggestions from interested parties about the review of TD/1, which should be available for industry comment by March 2000. Proposals, with any suggested solutions, should be addressed to Keith Nixon, secretary to PTD/1, Institution of Gas Engineers, 21 Portland Place, London, W1N 3WAF.

4. ACKNOWLEDGEMENTS

The authors would like to acknowledge the Institution of Gas Engineers for their kind support in the publication of this paper, together with contributions from members of the PTD/1 Panel.

5. REFERENCES

1. IGE/TD/1 Edition 3. 'Steel Pipelines for High Pressure Gas Transmission', Institution of Gas Engineers Recommendations on Transmission and Distribution Practice, 1993. Communication 1530.
2. prEN 1594. Gas Supply Systems - Pipelines for Maximum Operating Pressure Over 16 Bar - Functional Requirements. Final Draft. June 1999.
3. ISO CD 13623. Pipeline Transportation Systems for the Petroleum and Natural Gas Industries. March 1995.
4. IGE/SR/24. Risk Assessment Techniques. 1999. Communication 1655.
5. Oude Hengel, J, 'Limit State Design in Pipeline Codes', Conference on 'Risk and Reliability and Limit States in Pipeline Design and Operations, Aberdeen, UK, May 1997.
6. Lamb, M, Francis, A, Hopkins, P, 'How Do You Assess The Results Of A Limit State Based Pipeline Design?', Conference on Risk-Based and Limit State Design & Operation of Pipelines, Aberdeen, UK, October 1998
7. Jones, WP, Senior, G, Justifying The Uprating Of A Transmission Pipeline To A Stress Level Of Over 72% SMYS - An Operator's Experience, OMAE 99 conference, St John's, Newfoundland, July 99.
8. IGE/TD/1 Edition 3. Supplement 1. Handling, Transport and Storage of Steel Pipe, bends and Fittings, 1999. Communication 1657.
9. IGE/TD/1 Edition 3. Supplement 2. Design of 1219.2 mm (48 inch) Diameter Pipelines, 1999. Communication 1658.
10. IGE/TD/3. Edition 3. Distribution Mains, 1992. Communication 1514.
11. IGE/TD/9. Offtakes and Pressure-Regulating Installations for Inlet Pressures Between 7 and 100 Bar. 1986. Communication 1229.
12. IGE/TD/10. Pressure-Regulating Installations for Inlet Pressures Between 75 mbar and 7 Bar. 1986. Communication 1256.
13. IGE/TD/12. Pipework Stress Analysis for Gas Industry Plant. 1985. Communication 1252.

14. Francis, A & Senior G, 'The Use of Reliability based Limit State Methods in Uprating High Pressure Pipelines', International Pipeline Conference, Calgary, June, 1998
15. NEN 3650. Eisen Voor Stalen Transportleidingsystemen (Requirements for Steel Pipeline Transportation Systems). nederlands Normalisatie-Institut. Delft. 1992.
16. Leach G, Young AT, Transco. Assessing the Influence of Soil Loading and Restraint on In-Service Pipelines. IMechE Ageing Pipelines Conference, Newcastle, United Kingdom. 11th - 13th October.

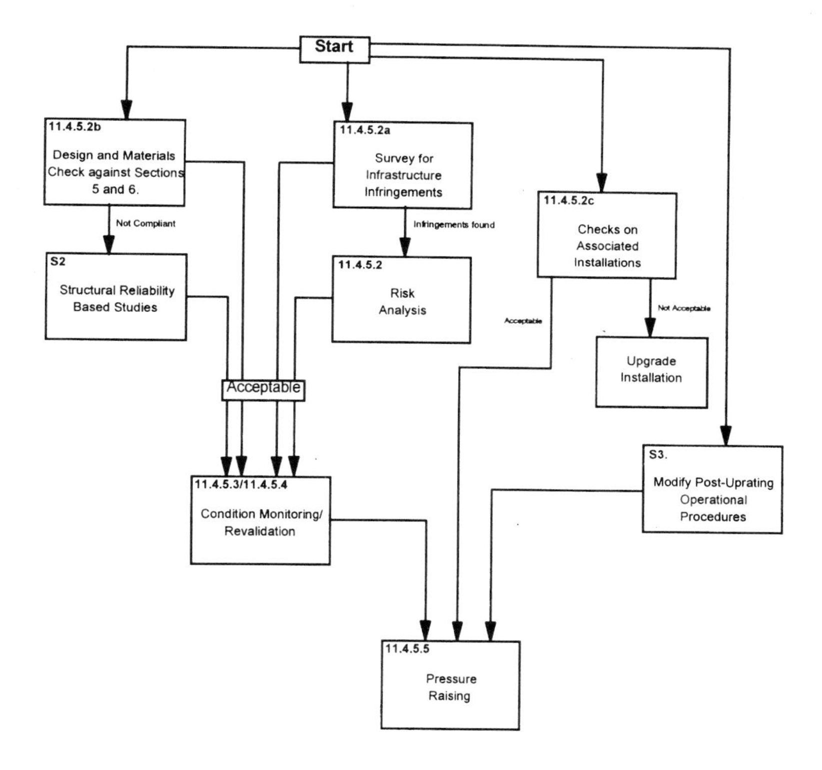

Figure 1 - Overview of Uprating Procedure

C571/028/99

Revalidation of pipelines beyond design life

R H FICKEN
Technical Services, Marine and Pipelines, Total Oil Marine plc, Aberdeen, UK

ABSTRACT

This paper discusses the experiences of Total Oil Marine plc in revalidating offshore trunk lines and flowlines beyond original design life.

The process of revalidation is described and lessons learned are highlighted.

The paper considers revalidation in relation to the responsibilities generated by the introduction of the Pipeline Safety Regulations, 1996. Since the demise of certification as a means of demonstrating safety, the purpose of revalidation has become unclear. The continued role of revalidation is discussed.

1 INTRODUCTION

Total Oil Marine plc (TOM) has revalidated both offshore trunk lines and flowlines beyond their original design life. Revalidation has been applied on a trap to trap basis and includes fittings. To date, TOM has revalidated both the 32" Frigg trunk lines and the infield 6" PN1 flowline.

The Frigg system exports gas from several fields. The two 32" Frigg lines extend 365 km from the Frigg complex to a shore-terminal at St. Fergus, in Scotland. The Frigg lines were installed during the years 1975 to 1977 and were trenched into the seabed. In October 1977, the lines were commissioned with a 20 year design life. The 6" PN1 flowline ties back a subsea wellhead to TOM's Alwyn North platform. The flowline was laid on the seabed and is protected beneath a rockdump along its entire 5.5 km length. Two trenched umbilicals provide chemical injection and control facilities for the wellhead. The 6" PN1 flowline was laid in May 1987 and commissioned in June 1988, with a design life of 10 years. A schematic of TOM's pipeline system is shown overleaf.

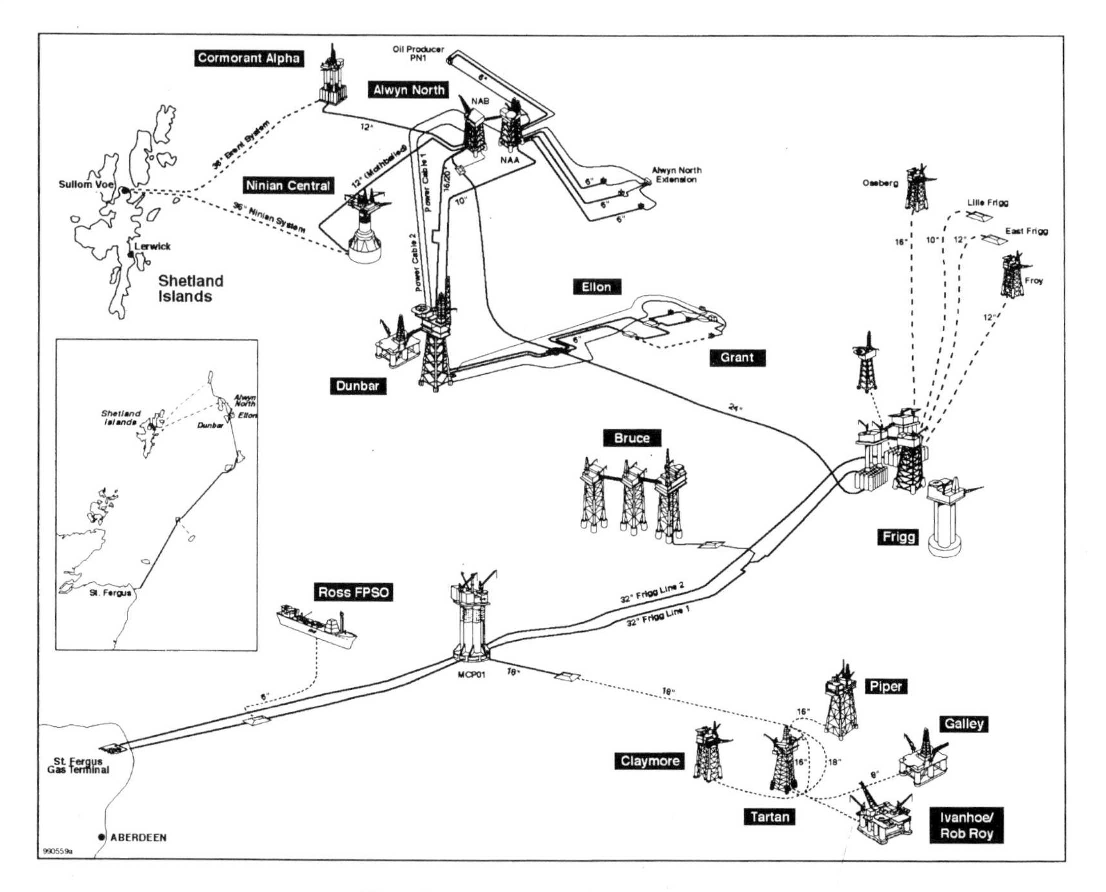

Figure 1 TOM pipeline system

Both the 32" Frigg lines and the 6" PN1 flowline remain in service despite having exceeded their original design life. Revalidation was sought for the period of their anticipated additional service lives. For both of the Frigg lines, the additional service life was nominally deemed to be for a further 20 years. For the PN1 flowline system the additional service life was set for a further 10 years. The cases for revalidation have been presented to the UK Health and Safety Executive (HSE) and the Norwegian Petroleum Directorate (NPD) without objection.

When TOM began the process of revalidating these lines, no guidance existed. A common sense approach was developed in consultation with the HSE and NPD. However, during the process of these revalidations, the régime for the management of pipeline safety was transformed. The introduction of the new Pipeline Safety Regulations, 1996 [(1)] (PSR) shifted the management of pipeline safety from a régime of prescription, to one of goal-setting. The emphasis of revalidation has necessarily changed.

This paper discusses revalidation in relation to the responsibilities generated by the introduction of the PSR, describes the approach used by TOM and highlights the lessons that have been learned while undertaking this process.

2 LEGISLATIVE ENVIRONMENT

Responsibility for the management of pipeline safety, prior to the introduction of the PSR, essentially rested with the Authorities. Safety was based on certification. Pipelines were deemed safe for their design life if they were designed to an acceptable code and the prescribed annual inspections were carried out. Prescriptive inspection would not necessarily inspect for all failure mechanisms and, therefore, there was an implied reliance on a good design code, that would also have to include some conservatisms, to catch those failure mechanisms that were not inspected for during the design life.

Under this system, the revalidation or 'recertification' process was significant for ensuring pipeline safety. Once the design life had expired, the basis for the management of pipeline safety was no longer valid. A recertification process would require consideration of the current utilisation of the conservatisms within a code for all those areas not assessed as part of the prescriptive inspections. This would require a full design review against the design code. If it was found that the pipeline continued to be acceptable to code, it would be possible to recertify the pipeline and continue managing safety as before.

However, fitness-for-purpose concepts had begun to enter pipeline maintenance strategies for the assessment of defects and failure mechanisms. Fitness-for purpose is not the same as being acceptable to design code. The concept of fitness-for-purpose allowed utilisation of some of the inherent conservatisms within design code. The problem associated with this was that these conservatisms were also being used to mitigate against the lack of comprehensive consideration of all failure mechanisms. The advantages of fitness-for-purpose assessment and the problems associated with its piecemeal implementation were addressed by the introduction of the new goal-setting PSR.

The PSR require that the operator is able to demonstrate that all their pipelines are fit-for-purpose, whenever required to do so. This implies a full review and assessment of all the failure mechanisms, consideration of time related effects, i.e. ageing, and updates whenever there is new information. As ageing is now taken into account, the question is raised as to the relevance of design life in the demonstration of pipeline safety.

TOM was undergoing the process of revalidation during the change in the legislative framework. Consequently a full design review had not previously been carried out. Such a review was required to ensure that all failure mechanisms had been addressed, enabling demonstration that all TOM's pipelines remained fit-for-purpose. Although the design review process is described in terms of revalidation, the process transformed into more specifically a justification for an extension to the design life as part of an ongoing demonstration of fitness-for-purpose. If a revalidation was undertaken today, a full and updated design review would already have been available, as required under the new legislative framework. Therefore, the revalidation process as described below, considers fitness-for-purpose issues that would now be dealt with elsewhere.

3 REVALIDATION PROCESS

Revalidation requires an evaluation of current fitness-for-purpose and a demonstration of future integrity. The process of revalidating a pipeline is essentially a design review to identify the design issues and to assess whether revalidation is appropriate. The results of a revalidation process may be that it is safe to continue to operate at the conditions indicated in the original design, or that a revision to the current operating conditions is required. A prediction of current remaining life is also generated.

3.1 Design review

The first major philosophical decision is whether to review an entire pipeline design or to restrict the review to those areas of the original design that have changed. TOM opted to review those areas of the original design that have changed. The DNV 'Rules for Submarine Pipeline Systems, 1996 [(2)] have adopted the full design review approach. Once having decided on a full review, there is a second decision as to whether to undertake the review against the original code or one that is more recent.

The DNV Rules, Section 11, 'Condition Assessment / Re-qualification,' require that the whole pipeline design is reviewed against the design section of the current DNV Rules. This applies to lifetime extensions beyond the minimum of 15% of the original design life or 5 years, except as otherwise approved by DNV. The exception is qualified by the requirement to demonstrate that an alternate requalification method maintains the same target safety level as the current DNV Rules.

The TOM approach assumes that the original design code ensured pipeline safety. This assumption is supported by the evidence gathered during the life of the pipeline and the lives of other pipelines designed under their original codes. If there have been no changes and no failures it appears strange to expend effort assessing an entire design. The DNV approach assumes that the original code may not necessarily provide the same target safety level as the current DNV Rules. The level of safety is generated by the code as a whole, rather than by individual elements within that code. In a revalidation, certain design elements are fixed by the fact that the line already exists. Therefore, when using different codes in a revalidation it is unlikely that the levels of safety are directly comparable. This could potentially lead to conservative revalidation assessments. Codes are also updated to take account of current manufacture, fabrication, installation and inspection techniques. Consequently there are questions relating to the application of new codes for the revalidation of pipelines designed under older codes.

The DNV Rules for short lifetime extensions, or changes to the design basis while within the original design life, are based on a restricted review of only those elements of the design that have changed. This section of the DNV Rules is very similar to the approach developed by TOM.

The approach developed by TOM is discussed in the following sections. The question of whether the TOM or DNV method of revalidation is best lies outwith the scope of this paper.

3.1.1 Identification of changes to the design conditions and issues raised

As part of the design review changes to the design considerations must be identified. Changes to the original design conditions are likely to have occurred and each change raises various issues that have to be assessed as part of the revalidation process. Changes may have occurred due to the following:

i. Errors / Unknowns at the design stage
ii. Revised prediction based on operational knowledge
iii. Extended design life

3.1.1.1 Errors / Unknowns at the design stage

During the design process, an attempt is made to consider all the issues that affect the design. However, errors will sometimes occur. These errors will usually become apparent during the design life of the pipeline and will raise issues that require to be assessed.

One such error occurred when designing the cathodic protection system for the PN1 flowline. Incorrect consideration was given to the pipeline operating temperature and its effect on the required pipeline current densities and anode capacity. This reduced the life of the original cathodic protection system to approximately 3 years.

The designer is unable to mitigate against new and as yet unknown design issues. Such issues are usually associated with the application of new technologies and materials. This was the case with the PN1 flowline and the then new coating material, ethylene propylene diene monomer (EPDM). The EPDM had passed the standard bench testing for electrical resistance, however, it was later found that under hydrostatic pressure this early formulation of EPDM was both semi-conductive and cathodic to carbon steel. The standard bench testing had been inadequate for the assessment of this coating as the flowline was installed subsea.

3.1.1.2 Revised design predictions based on operational knowledge

Certain predictions have to be made at the design stage. These predictions are characterised by a scarcity of initial data. During the design life of a pipeline real data may be gathered to enable revised predictions to be made. Predictions at design often have to be made for future fluid conditions, environmental conditions and external loadings.

Predictions for produced fluids would initially have to be based on the well tests. Fluid flow, pressure, composition and lifetime predictions for trunk lines would usually be based on initial well tests for the initial feeder fields and include assumptions regarding fluids from future tie-ins. Clearly, these predictions cannot be particularly accurate and, indeed, it is the inaccuracy of the lifetime predictions that have required revalidation beyond the initial design life.

Good data on environmental conditions is difficult and expensive to collect. The range of environmental conditions are only exhibited over a long period and often vary over the length of a pipeline. Usually during the life of a pipeline, there is an opportunity to improve on the dataset used for the prediction of environmental conditions.

During the life of the PN1 flowline site specific environmental data has been collected on the Alwyn North platform allowing the initial predictions to be re-assessed. The most significant results were: a reduction in the 100 year storm wave; an overall increase in the number of wave cycles experienced, but a significant reduction in the waves over 4 metres high; much lower current velocities than originally predicted; and shorter storm wave periods, i.e. steeper waves than previously predicted. The revised wave and current data have affected the severity of the loads imposed on the pipeline system and the change to the wave spectrum has modified the number of cycles that have to considered in the fatigue assessments.

Many of the pipelines, that are now requiring revalidation, were designed with the most limited environmental datasets. These older pipelines are those that tended to be installed at the beginning of development within a region, when few surveys had previously been conducted. This was true of the Frigg lines, installed at the beginning of development in the North Sea. Investigations indicate that the original dataset yielded only slightly conservative predictions.

Changes to other external loadings are less obvious and are generally difficult to predict. For example, the recent Trenching Guidelines JIP [(3)] has highlighted the constantly changing patterns of the loadings associated with fishing gear interaction. New techniques, gears and varying intensities give rise to loadings that are difficult to predict and may have varied considerably from the design. The changes to loadings associated with the weight of drill cuttings on the base of the PN1 riser provide a further example. These loadings are typically of the type that would not have been captured under the prescriptive régime. In this case, the review found that the height considered in the design had been exceeded and the issue of possible riser over-stress had to be assessed.

3.1.1.3 Extended design life

Some failure mechanisms are time dependent. A pipeline system is usually designed with a finite life based on anticipated field life. Reassessment of time dependent failure mechanisms is necessary when there is an extension to the predicted service life. Operational knowledge and trending of inspection data may allow possible conservatively predicted degradation rates and safety factors to be replaced by real data and realistic predictions of future degradation.

Degradation due to fatigue and the life of sacrificial cathodic protection systems are clearly time dependent. Reassessments of remaining life are required when there is an extension to the original design life. Updated environmental data is often available for enhancing the prediction of fatigue. Designs for sacrificial cathodic protection systems are often conservative and surveys are likely to show that remaining life will be available. Degradation of some materials is also time dependent. Polymers tend to degrade quite rapidly and, therefore, the extension to the design life of the PN1 flowline system required a reassessment of the remaining umbilical life.

3.1.2 Data requirements

Gathering of data is central to the revalidation process. Details of the original design and pipe history must be retrieved to identify the changes to the design that require to be assessed. The data search must be sufficiently wide ranging to seek out information that already exists on those failure modes that would not normally have been checked for under the previous prescriptive legislation. Some data will not be available and additional inspections will be required at revalidation. Additional baseline surveys are also necessary, since predictions become increasingly unreliable with time.

The wide ranging data search for the PN1 flowline uncovered information that had not previously been considered. The drill cutting survey, performed as part of the Alwyn North platform structural assessment, also proved useful for the assessment of loadings associated with the weight of the cuttings on the riser base. The ROV survey of the Frigg lines was considerably extended. Increasing the sample of the condition of the Frigg lines enhanced confidence that all failure mechanisms had been considered. For both the Frigg lines and the PN1 flowline, prediction of corrosion had to be confirmed. Internal metal loss surveys enabled new baselines to be established. The surveys also provided additional data points for trending degradation rates into the future.

3.1.3 Revalidation assessment

Once the issues relating to the changes to the original design have been identified, these have to be assessed. Standard fitness-for-purpose assessments and prediction of future life techniques are used. TOM undertook a number of individual assessments for a range of failure modes, using in-house engineering and variety of specialist consultants. A summary report was required to collate these individual studies and present them as a coherent revalidation document.

4 LESSONS LEARNED

The revalidation process has not been straightforward. Various lessons have been learned that will benefit the future management of TOM's pipelines.

4.1 Time / Resource considerations

The amount of work and time involved in undertaking a revalidation exercise is considerable. Effective planning is required to achieve revalidation before the design life for a pipeline has expired. On the Frigg lines, TOM began work three years prior to revalidation.

Retrieval of design data and the compilation of a pipe history from archived records takes a considerable amount of time. Planning and execution of specialist surveys, such as internal inspections, can only begin once the requirement has been identified. These can be expensive and time consuming. For example, an internal survey can rarely be completed from requirement to results in less than a year. Where, as in the case for PN1, a suitable tool did not exist and development was required, the time scale can be considerable. The length and complexity of the assessment stage depends on the results. Much effort was expended on a necessarily sophisticated corrosion assessment for the Frigg lines. It should be noted that a revalidation process is likely to require many fitness-for-purpose studies addressing various failure mechanisms.

4.2 Archiving
Many older pipelines, such as Frigg, were installed prior to the widespread use of computers. Archived data is paper based and lay records tend to be handwritten. Consequently data handling and retrieval is difficult. Data for later pipelines, such as the PN1 flowline, was often generated electronically, however, the importance of electronic storage had not yet been recognised and hardcopy was archived. This remains a problem within the industry and electronic archiving has yet to be fully addressed.

Another archiving problem is that changes that affect pipelines, defect assessments etc. have rarely been associated with the pipeline design data. It is therefore difficult to reconstruct pipeline histories.

4.3 Data acquisition
When undertaking fitness-for-purpose reviews it becomes clear that certain data would be very useful if only it had been recorded. This is very true of process data. Real pressure, temperature, flowrate and compositional information could be used in fatigue assessment, corrosion prediction and many other pipeline fitness-for-purpose and integrity prediction applications. However, process applications tend only to require a memory of a few hours and then the information is over-written. When data is recorded, again it is often not targeted at the pipeline engineer. Daily averages or spot readings may mask fluctuations that can be important to the calculations. Much data, that should be easy and inexpensive to archive, is discarded.

The design of many pipelines ensures the acquisition of data is difficult. For example, the PN1 riser is of a smaller bore than the flowline. An internal inspection tool could not fit through the riser and an expensive subsea intervention had to be undertaken for the line to be surveyed. Although the PN1 flowline was never intended to be pigged, this continues the theme that the data requirements for pipeline assessments are rarely considered. This is partly due to the fact that demonstration of fitness-for-purpose has only been a requirement since the introduction of the PSR, but must now be addressed.

4.4 Data reconciliation
Much defect assessment work requires comparing and combining different datasets. The difficulty of this task is frequently under-estimated.

As an example, an internal survey may detect a defect at a certain position located by an odometer wheel and a weld count. From the location it should be possible to use the lay record to determine which pipe joint the defect is within. The pipe joint details should lead to material certificates enabling real material properties to be used in any assessment. Further to this the location should enable cross-referencing of external survey data from an ROV for example. This would allow for a diver to undertake a more detailed inspection or for the possible assessment of additional bending stresses introduced by freespanning.

In the real world the odometer wheel has a limited accuracy and internal inspections cannot guarantee to count the correct number of welds. Lay records for old pipelines are notoriously incomplete. Hand-written lay records, such as those for Frigg, by their nature, include errors and are particularly difficult to interrogate. The use of pipeline kilometer-post or KP to define external pipeline locations is also problematic.

The KP generated during a survey is related to a KP generated on the laybarge and recorded on the lay record. Unfortunately, KP's on the lay record are rarely the same as those generated during a survey. Laybarge KP's often represent the linear length of a pipeline as it is laid. Survey KP is usually the length as a distance over a reference spheroid and will generate shorter apparent lengths when a pipeline encounters a slope. Where there is as-laid data that can be cross-referenced to the lay records, problems remain. As-laid surveys for old pipelines used old and inaccurate survey and positioning systems with low repeatabilities. Many survey databases have since been refined with modern positioning systems, however, if this is done the lay record must also be updated. Modern positioning systems may have the repeatability to return to within the same pipe joint, however, the KP is unlikely to relate to a uniform reference.

Data reconciliation is therefore very difficult. On short unburied lines it may be possible to count welds externally from one end and generate a new pipe tally. On long and buried lines, such as Frigg, the operator has to resort to the placing of local reference markers that can be clearly identified from within and outwith a pipeline. The problem can be solved on new lines by specifying that the as-laid survey reports the position of each girth weld as these can be seen by internal and external surveys and are a sufficient distance apart that modern survey accuracies will be able to distinguish between them.

For older lines uncertainty will remain. Each piece of data has to be regarded with an error band. When using this data to combine loads worst case combinations have to be considered. This is extremely time consuming on a paper based system and is potentially very conservative. To mitigate against these problems TOM has begun to convert data for all of its pipelines into a relational database format that will enable rapid reconciliation of data.

5 THE FUTURE FOR REVALIDATION

The revalidation process undertaken by TOM includes consideration of the fitness-for-purpose of those areas that had changed from the original design. As discussed earlier, this aspect has become an ongoing requirement under the PSR and would not now be required in the context of a revalidation. However, it could be argued that continuous revalidation is required for the demonstration of pipeline safety.

In the era of 'recertification' the time interval at which to revalidate a pipeline would have been the design life. Now revalidation must be continuous. In reality the PSR implies that revalidations are controlled by risk based systems that determine when a potential failure mode requires to be assessed and when to inspect for each of these failure modes.

Ageing has come to be considered within the ongoing assessment of pipeline integrity. The safety aspect of a defined design life is no longer relevant. However, remaining life will continue to be of relevance for commercial considerations, if not for safety. As the word 'revalidation' implies safety, it is probable that future life prediction will become separated from that term. Future life prediction is likely to be provided as a module within the risk based systems that provide the demonstration of pipeline safety.

6. SUMMARY

Demonstration of pipeline safety through 'recertification' is obsolete and revalidation has now become an integral part of the ongoing demonstration of pipeline safety. The limitations of prescriptive legislation and the advantages of the new goal-setting régime have been discussed. TOM's revalidation experiences during the introduction of the new legislation provide a useful insight into the concepts behind the change. The experience also allowed TOM to gain a headstart and learn early lessons in fitness-for-purpose based demonstration of pipeline safety. This is now required on an ongoing basis for all pipelines.

REFERENCES

1. Statutory Instruments No. 825 (1996) 'Pipeline Safety Regulations'
2. DNV Rules for Submarine Pipeline Systems, 1996
3. The Trenching Guidelines JIP, Trevor Jee Associates, 1998

Ownership note
TOM operate the Frigg Lines and the PN1 flowline on the behalf of various owners. The ownership consortium for Frigg line 1 includes TOM (33.33%) and ELF Exploration UK PLC (66.66%). The Frigg Line 2 ownership consortium includes Elf Petroleum Norge A.S. (21.42%), Den norske stats oljeselskap a.s. (29.00%), Total Norge A.S. (16.71%) and Norsk Hydro ASA (32.87%). The PN1 flowline is owned by TOM (33.33%)and Elf Exploration UK PLC (66.66%).

C571/032/99

Recent trends in gas pipeline incidents (1970–1997): a report by the European Gas pipeline Incidents data Group (EGIG)

R BOLT
Safety Department, N V Nederlandse Gasunie, Groningen, The Netherlands
R W OWEN
Pipeline Services, BG Technology, Loughborough, UK

SYNOPSIS/ABSTRACT

The European Gas pipeline Incident data Group (EGIG) gathers data on the unintentional release of gas in their pipeline transmission systems. The EGIG data presents the best available failure statistics for Western Europe gas transmission pipelines, and so is able to demonstrate their high level of safety. The paper presents information on the development of the database and accumulated frequencies, including the total incident frequency over the 1970 to 1997 period, the experience over the last five years, and the main contributing causes of failure.

1. INTRODUCTION

In 1982, six European gas transmission system operators took the initiative to gather data on the unintentional gas release of gas in their pipeline transmission systems. This co-operation was formalised by the setting up of the European Gas pipeline Incident data Group (EGIG).

The objective of the group is to provide a broad basis for statistical use, to give a more realistic picture of the frequencies of incidents than would be possible with the independent data of each company considered separately. The collection of safety related data has grown in significance as a result of the increasing interest shown by local, national and international authorities responsible for safe gas transmission. The EGIG data, therefore, represents the best available failure statistics for Western Europe gas transmission pipelines, and so is able to demonstrate their high level of safety. The EGIG data also represents the possibility of "benchmarking" against other pipeline operators' experience.

In 1997, a total of nine companies were participating in EGIG, comprising all of the major gas transmission system operators in Western Europe. The participating companies were :

- Dansk Gasteknisk Center a/s, represented by DONG;
- ENAGAS, S.A.;
- Gaz de France;
- N.V. Nederlandse Gasunie;
- Ruhrgas AG;
- S.A. Distrigaz;
- SNAM S.p.A;
- SWISSGAS AG;
- Transco, represented by BG Technology.

In setting up the data collection scheme, two aspects required particular attention. Firstly, it was necessary to obtain an accurate database of pipeline exposure (kilometre-years) for each size of pipeline in each country. This exposure information was derived for each year since 1970 for each company. Secondly, the incidents were collated for each year using a reliable and consistent definition, which did not depend on the individual companies' surveillance and inspection policies. Thus, the criterion of loss of gas was chosen for this purpose. This enables accurate data on the main parameters of exposure and failure to be obtained, thus leading to reliable values of failure frequency.

Considering the number of participants, the extent of the pipeline systems and the exposure period involved (from 1970 onwards for most of the companies), the EGIG database is a valuable and reliable source of information. The database contains a total of accumulative exposure of approximately 2 million kilometre-years gas transmission pipelines (up to the end of 1997), and the total number of incidents in the EGIG database is approximately 1000. The regional differences are not taken into account so that the database represents an average of all participating companies. Uniform definitions have been used consistently over the entire period. Consequently, the database gives useful information about trends, which have developed over the years.

In this paper, information is given on the development of the database and results of some analyses. Accumulated frequencies are given over the entire period. As the exposure time over the pipeline system increases, each new year added has a smaller effect on the accumulated frequencies. Therefore, a separate analysis of the most recent period is also given. This is done by using the 5-years moving average or by comparison of only the past five years with the accumulated frequency.

2. CLASSIFICATION OF INCIDENTS AND DEFINITIONS

2.1 Criteria for incidents

The criteria for the incidents in the database are :
- there is always an unintentional release of gas;
- the incidents are always related to an onshore steel gas transmission pipeline
    - with a design pressure greater than 15 bar;
    - outside the fences of installations;
    - excluding associated equipment (eg valves) or parts other than the pipeline itself.

2.2 Categories of leak size

Incidents are recorded into three categories, depending on the leak size :
- pinhole/crack : diameter of equal or less than 2 cm;
- hole : diameter of defect more than 2 cm and equal or less than the diameter of the pipe;
- rupture : diameter of defect more than the pipe diameter.

2.3 Types of incidents

The incidents are divided according to the initial cause into the following types :
- external interference;
- corrosion;
- construction defect/material failure;
- hot-tap made by error;
- ground movement;
- other and unknown causes.

2.3.1 Additional information

Depending on the type of the incident, the following secondary information is recorded, although this is only to explain possible differences in performances :

External interference :
- activity causing incident (eg digging, piling, ground works);
- equipment causing the incident (eg anchor, bulldozer, excavator, plough);
- installed protective measures (eg slabbing, casing, sleeves).

Corrosion :
- location (eg external, internal, unknown);
- corrosion type (eg galvanic, pitting, stress corrosion cracking, unknown).

Construction defect/material failure :
- type of defect (construction or material);
- defect specification (eg hardspot, lamination, material, field weld, unknown);
- pipeline type (eg straight, field bend, factory bend).

Ground movement :
- type of ground movement (eg dike break, erosion, flood, landslide, mining, river, unknown).

Other/unknown :
- causes are also subdivided into a number of pre-defined sub-causes (eg design error, erosion, lightning, maintenance, other weld, repair clamp, other/unknown).

For all incidents, other information is recorded. Some examples are : depth of cover, size of leak, type of leak (pinhole crack, hole, rupture, unknown), ignition (yes/no), detection (eg client, contractor, landowner, patrol), diameter, wall thickness, grade of material, construction year, design pressure, type of coating (eg asphalt, bitumen, coaltar, epoxy, polyethylene), other information (free text).

2.4 Failure frequency calculation

The failure frequency is calculated by dividing the number of incidents by the "kilometre-years", ie the exposed length for the pipeline category under consideration and its exposure duration. All the frequency figures are given per 1000 kilometre-years (kmyr).

3. DATABASE CONTENT

3.1 Development of database

The total length of the pipeline system of all the participating companies (up to the end of the 1997) is 105,919 km. The exposure in the 1970 to 1997 period is 1.98 million kilometre-years. The development of the database, from 1970 to 1997, is given in Figure 1.

3.2 Number of incidents

The number of incidents in the EGIG database is 945 up to the end of 1997. The annual number of incidents is presented in Figure 2.

4. RESULTS

4.1 Overall incident frequency

The development of the overall incident frequency is given in Table 1.

Table 1 : Overall incident frequency

Period	Number of incidents	Total exposure kmyr	Frequency (incidents per 1000 kmyr)
1970-1992	830	1.47 x 10E06	0.575
1970-1997	945	1.98 x 10E06	0.476

Compared to five years ago, the overall average failure frequency has decreased from 0.575 to 0.476 incidents per year per 1000 km pipeline. This represents a decrease of 17%.

An overview of the development of this average failure frequency over the total period 1970 to 1997 is presented in Figure 3. This figure shows the gradual reduction in the incident frequency in each year, ie the cumulative total from 1970 onwards.

In order to see the results over the most recent period for each year, a moving average has been calculated over the past five years (ie 1970 to 1974, 1971 to 1975 etc) and is also presented in Figure 3. The total incident frequency over the period 1970 to 1997 is 0.476 incidents per 1000 kmyr, but considering the last five years only (1993 to 1997), the incident frequency is much lower, 0.215 incidents per 1000 kmyr.

4.2 Frequency per type of incident

The frequency per type of incident (ie main causes) from 1970 onwards is presented in Figure 4. External interference represents 50% of the incidents, followed by construction defects/material failures (18%) and corrosion (15%).

Figure 5 shows the frequency per type of incident (initiating cause) over the total period (1970 to 1997) and also the performance over the last five years (1993 to 1997). Note the figures for 1993 to 1997 relate to that particular period and are not cumulative figures from 1970 onwards. This again shows that external interference remains the main cause for gas leakage incidents, but also shows that a significant improvement in the incident frequency has been observed in recent years.

4.3 Frequency by cause and type of leak

An overview of the incident frequencies by cause and type of leak in the period 1970 to 1997 is presented in Figure 6. This shows that the majority of pipeline ruptures have arisen from external interference, and that practically all loss of gas incidents due to corrosion have been pinhole cracks.

4.4. External interference

The frequencies for external interference for each range of diameter and type of leak are presented in Figure 7. This is not necessarily due to the diameter itself, but follows from the connection between diameter and wall thickness (ie larger diameter pipes tend to be of a thicker wall). The frequencies caused by external interference for each range of wall thickness and type of leak are presented in Figure 8. No external interference incidents resulting in gas leakage have been reported for a pipeline with a wall thickness greater than 15 mm.

An overview of the development of the average failure frequency due to external interference over the total period 1970 to 1997 is presented in Figure 9. This figure shows the gradual reduction in the overall incident frequency in each year, due to external interference, ie the cumulative total from 1970 onwards. In order to see the results over the most recent period for each year, a moving average has been calculated over the past five years (ie 1970 to 1974, 1971 to 1975 etc) and is also presented in Figure 9. This demonstrates the improvement in the incident frequency in recent years.

4.5 Detection of defects

The detection of incidents by type of detection is presented in Figure 10. More than 40% of all incidents in the EGIG database are detected by the public.

4.6 Ignition probability

On average, 3.9% of all of the incidents result in an ignition. The ignition frequency is related to the type of leak in Table 2.

Table 2 : Ignition probability

Type of leak	Ignition frequency %
Pinhole-crack	3.4
Hole	1.8
Rupture <=16 inches	7.5
Rupture > 16 inches	21.1

Although the frequency of an ignition is significantly higher for a rupture, the frequency of a rupture is much lower.

5. CONCLUSIONS

The European Gas Incident data Group (EGIG) collects incidents with gas release from high pressure onshore natural gas pipelines reported by nine European major gas transmission companies. The accumulated exposure of their pipeline system from 1970 up to 1997 is 1.98 million kilometre-years.

The EGIG database shows that :

- The total incident frequency over the period 1970 to 1997 is 0.476 incidents per 1000 kmyr;
- Considering the last five years only (1993 to 1997), the incident frequency is much lower, 0.215 incidents per 1000 kmyr;
- The reduced frequency over the last five years shows that the measures taken as a result of the lessons learned from recorded incidents and from the use of new technologies to inspect and maintain pipelines have been effective;
- Over the past decade, the overall frequency of incidents causing an unintentional gas release has gradually reduced demonstrating the success of an increasing integration of safety in the total pipeline process : improved design and construction (including pipe manufacture and pipeline construction), adequate maintenance, and safe operation;
- External interference remains the main cause of gas leakage with an average of 0.239 incidents per 1000 kmyr over the period 1970 to 1997. However, a improvement in the incident frequency has been observed in recent years, 0.103 incidents per 1000 kmyr over the period 1993 to 1997;
- External interference represents 50% of the incidents, followed by construction defects/material failures (18%) and corrosion (15%);
- In only a small minority of incidents did the leaked gas lead to ignition (3.9% on average);
- The levels and trends of incident frequencies go to demonstrate that the current national legislation and industry practices are adequate in ensuring safety, reliability and integrity of the natural gas pipeline systems;
- The use of pipelines for the transport of large quantities of natural gas to industry and to commercial and domestic consumers represents a safe mode of transport in terms of the impact on the environment and human health.

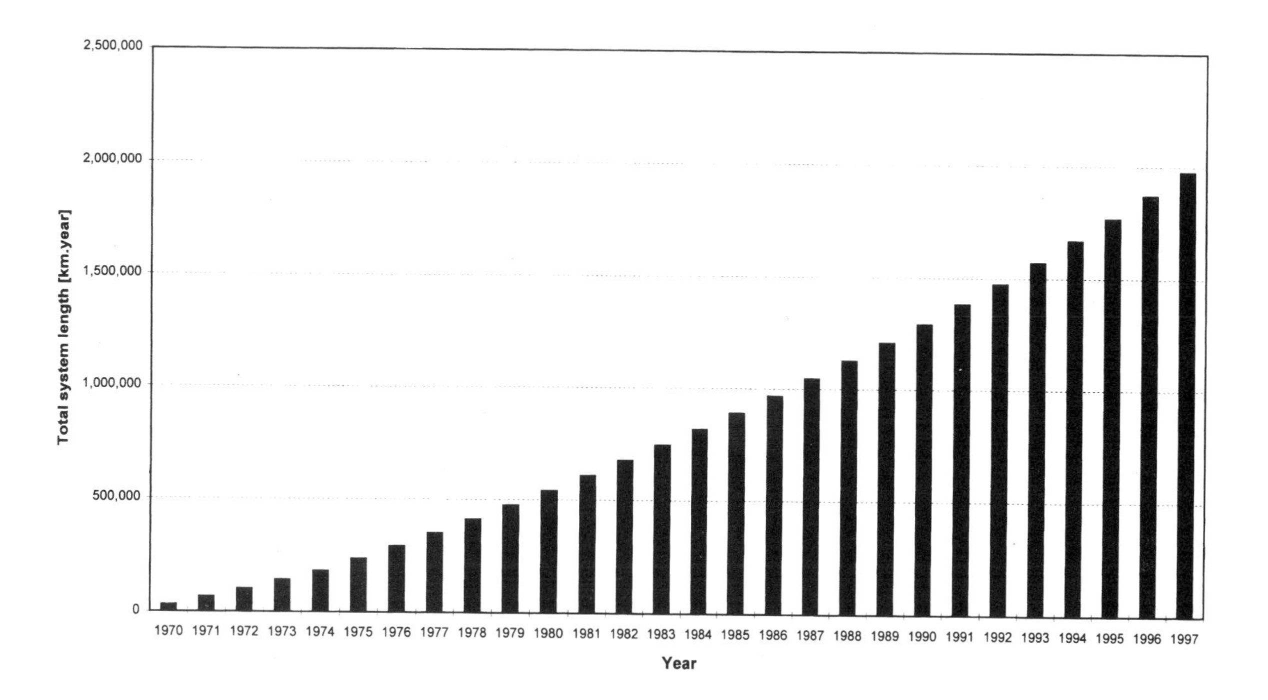

Figure 1 : Total growth of the database - cumulative system ~~length~~ exposure

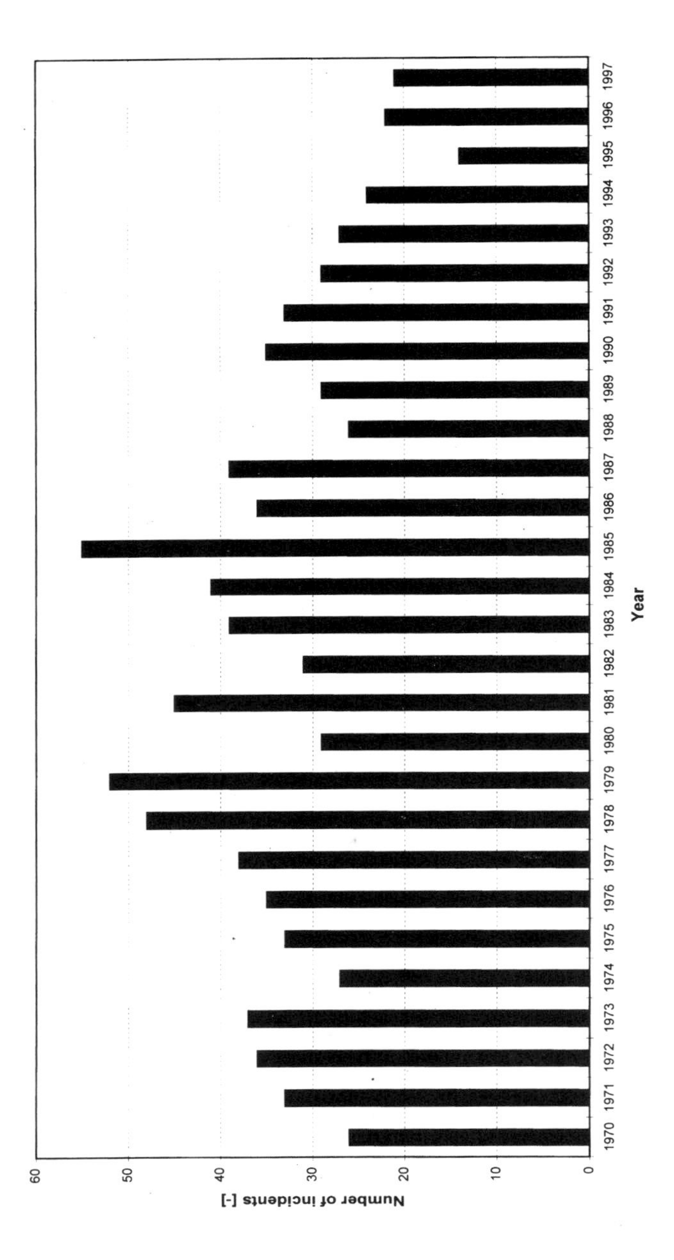

Figure 2 : Annual growth of database - number of incidents

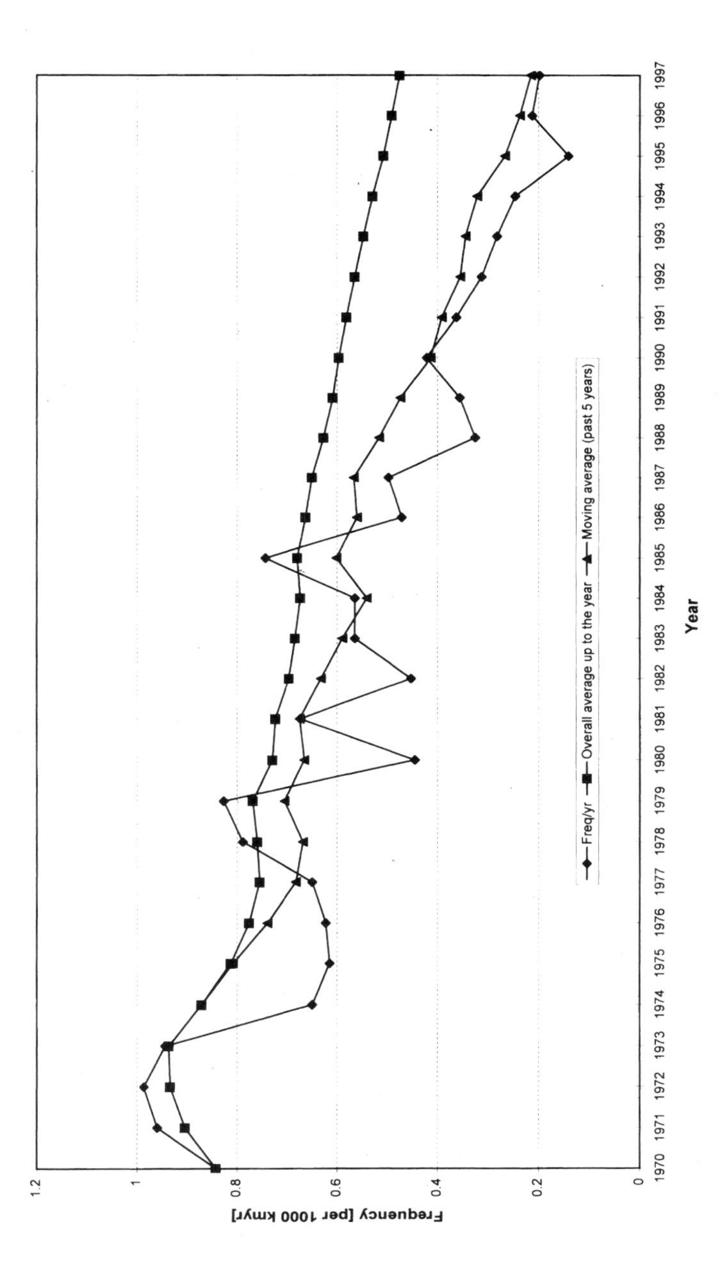

Figure 3 : Development of overall frequency

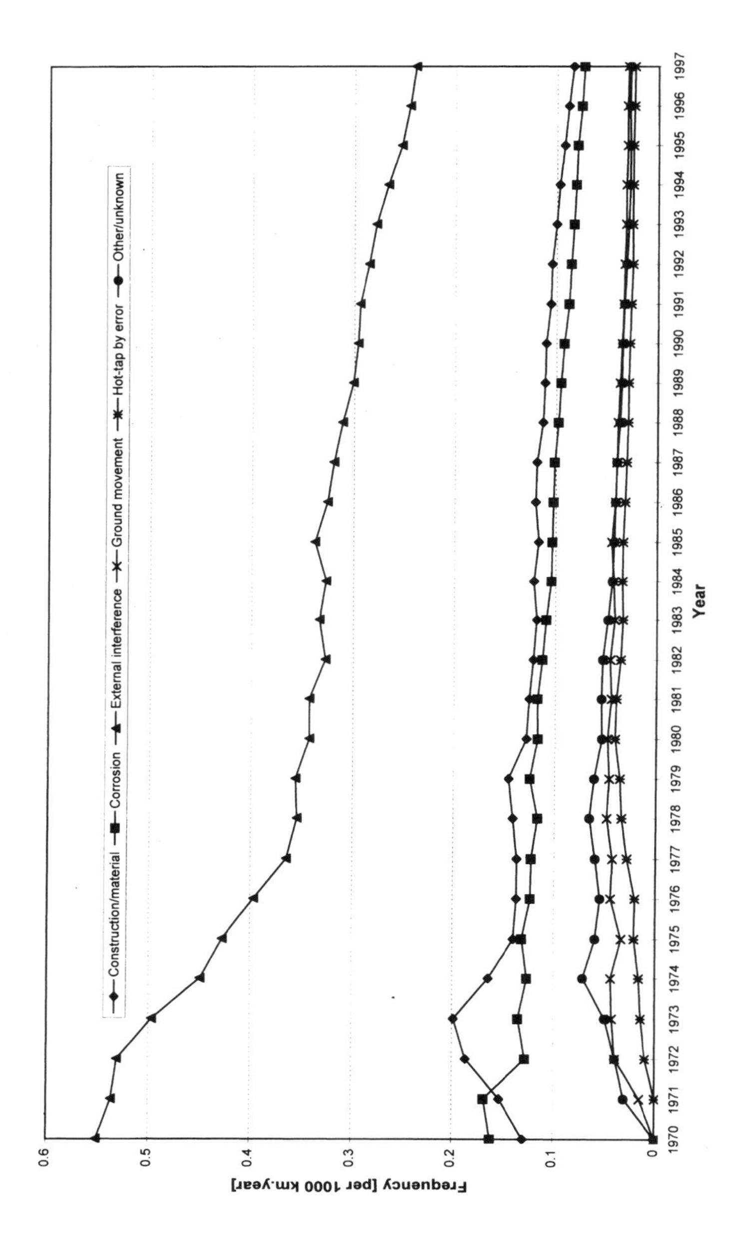

Figure 4 : Frequency per type of incident (cumulative) - main causes

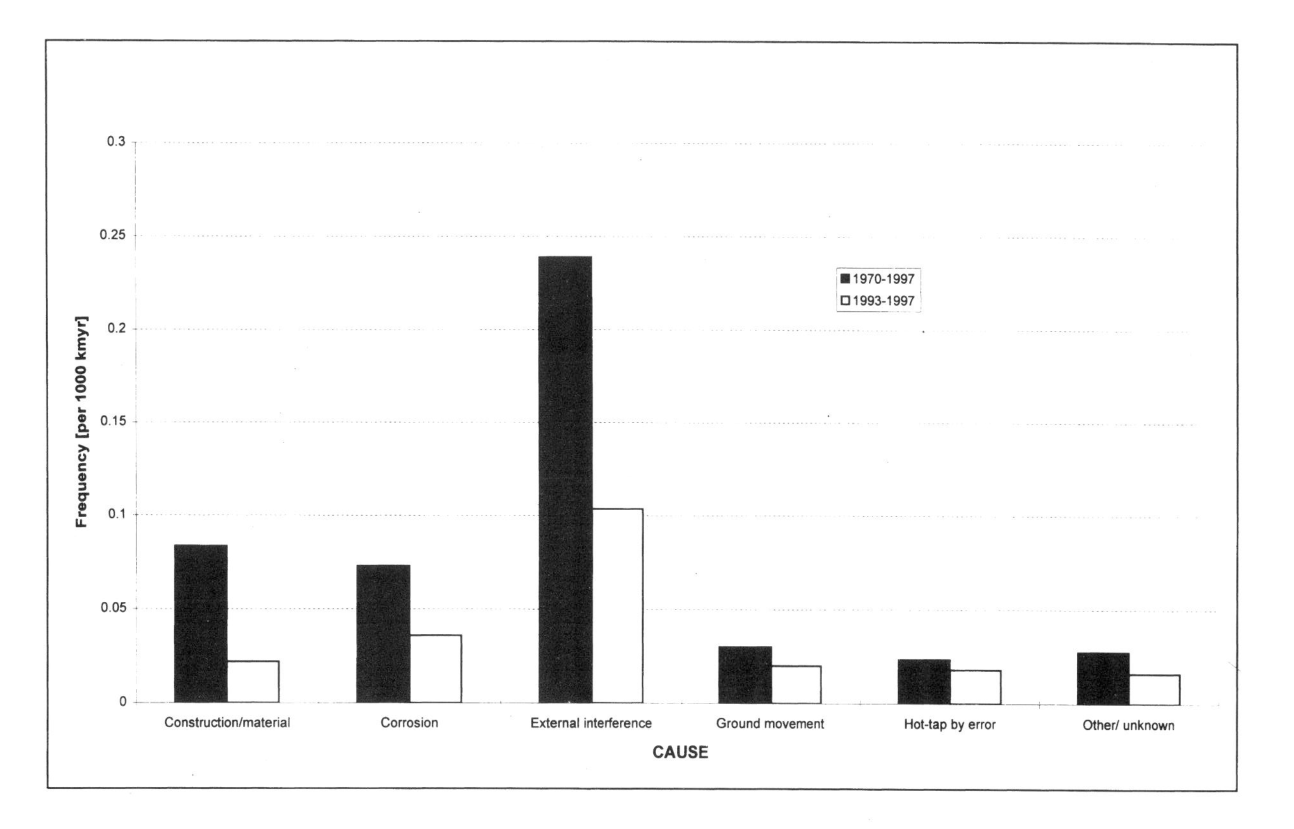

Figure 5 : Incidents by cause (cumulative) - long and short term period

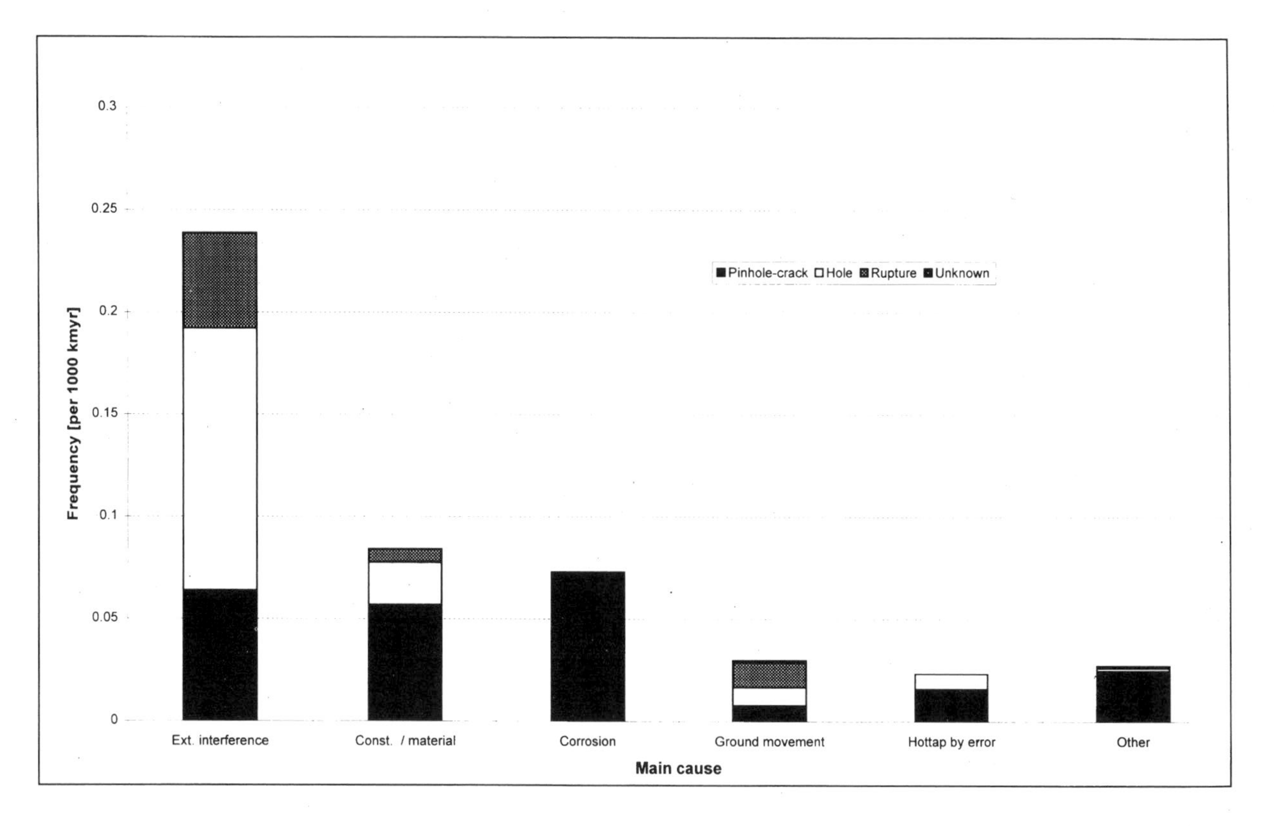

Figure 6 : Incidents by cause and type of leak

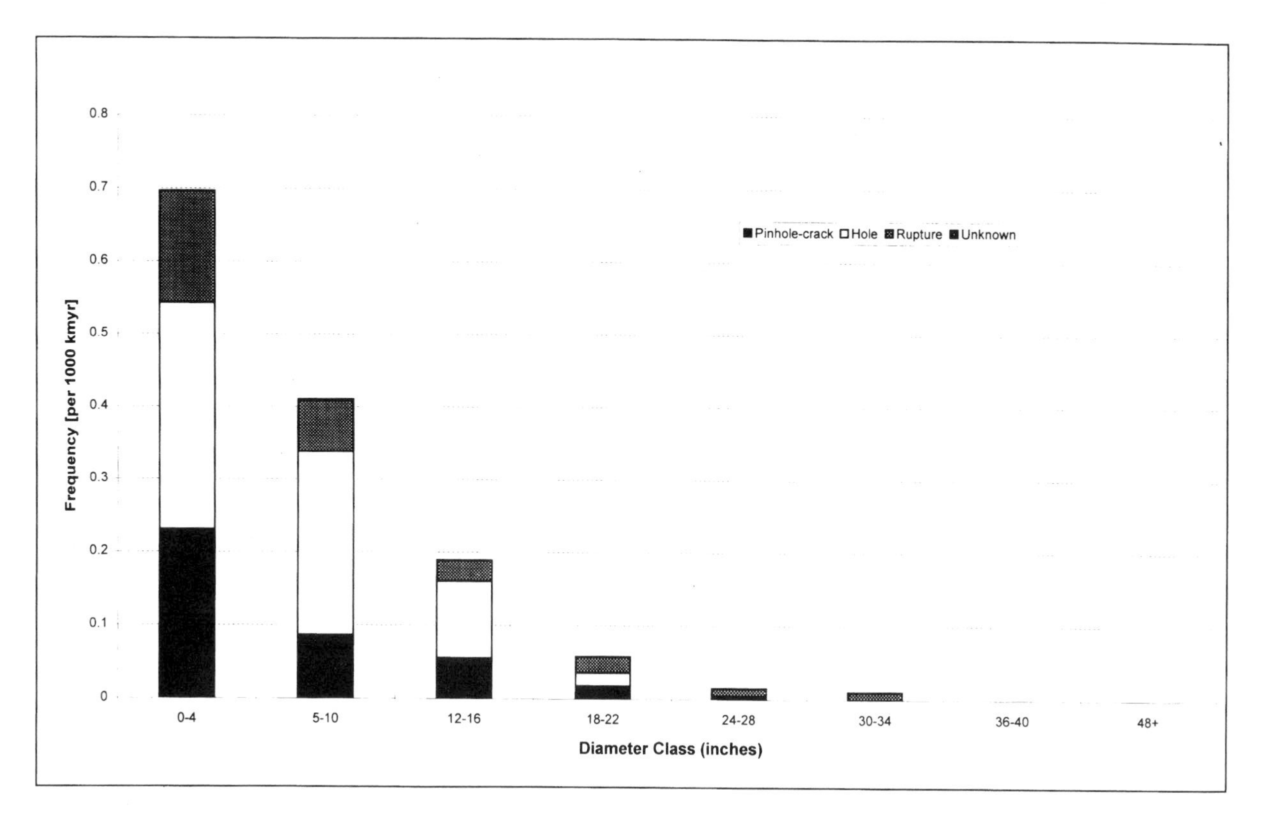

Figure 7 : External interference - frequency per diameter class

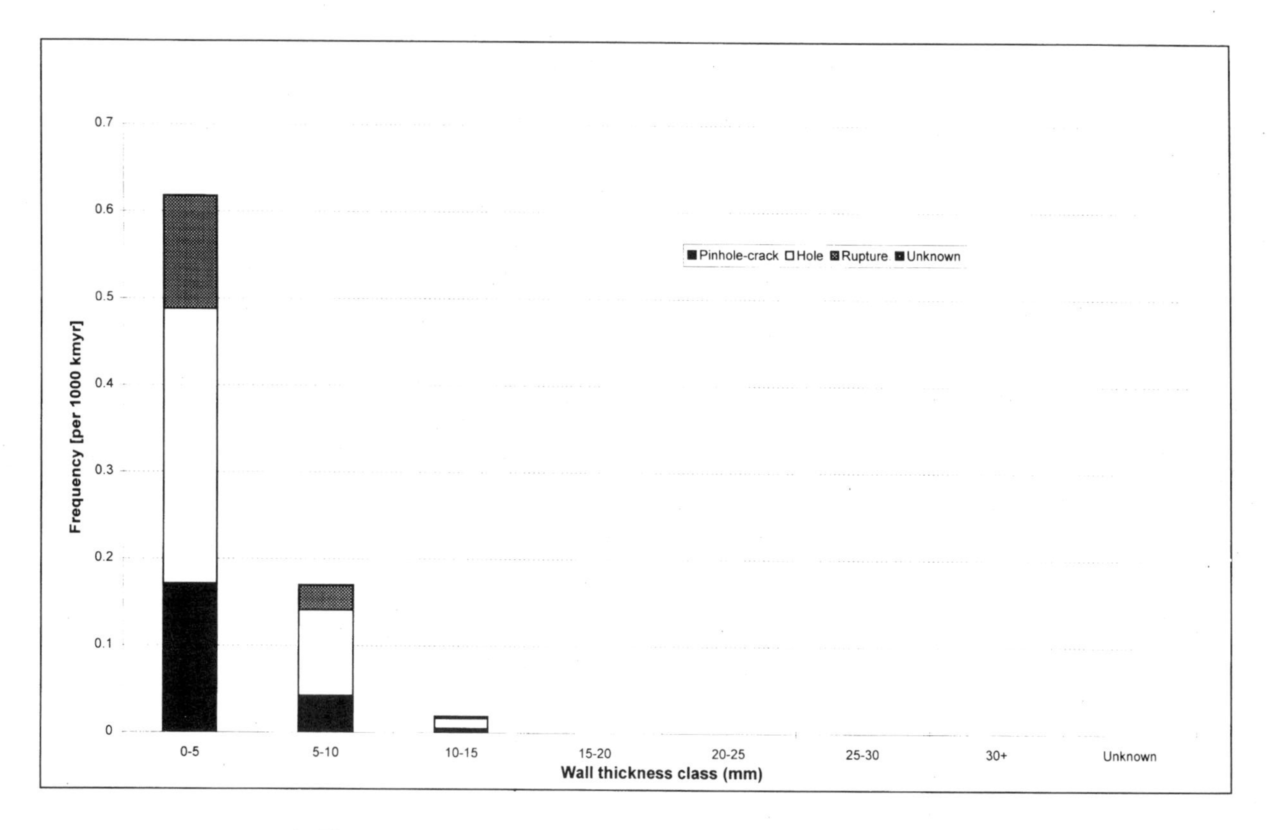

Figure 8 : External interference - frequency per wall thickness class

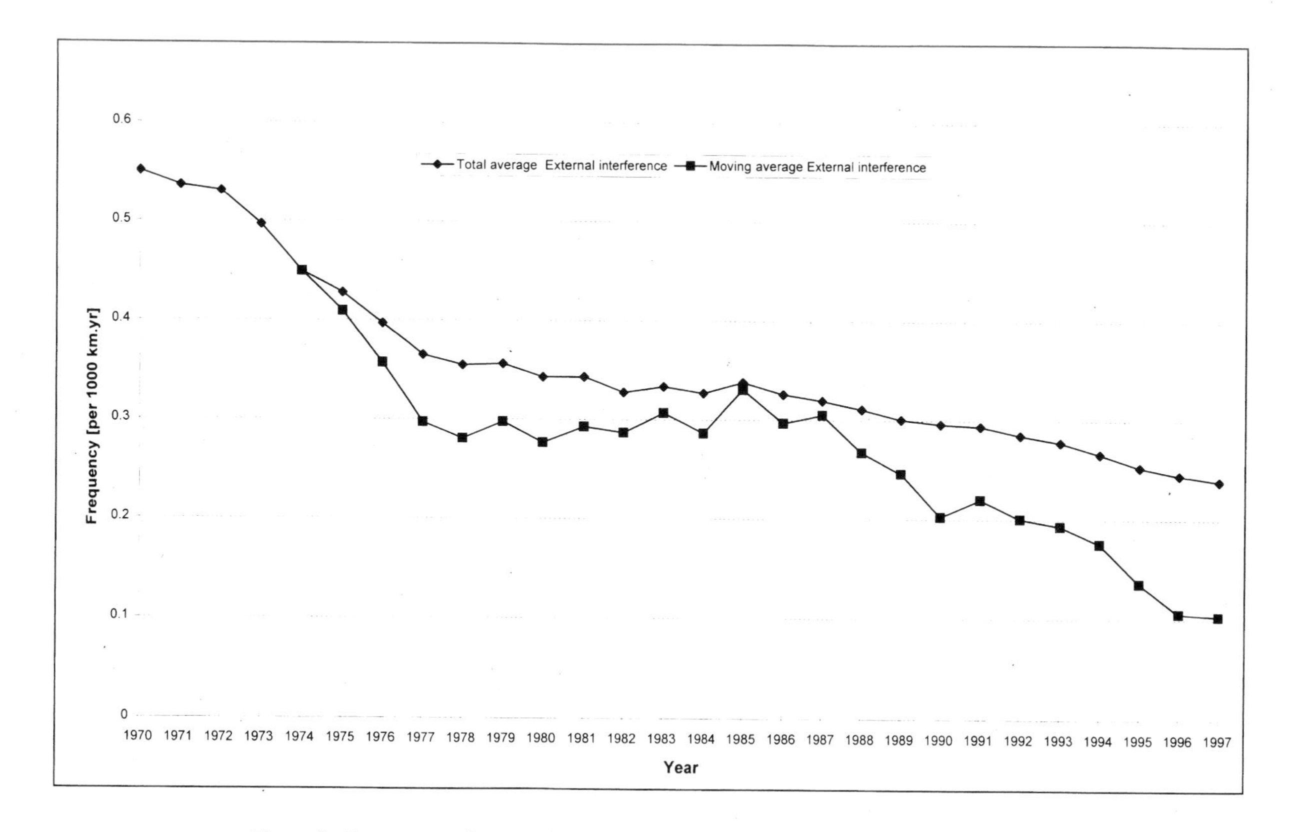

Figure 9 : Frequency of external interference - cumulative and 5 year moving average

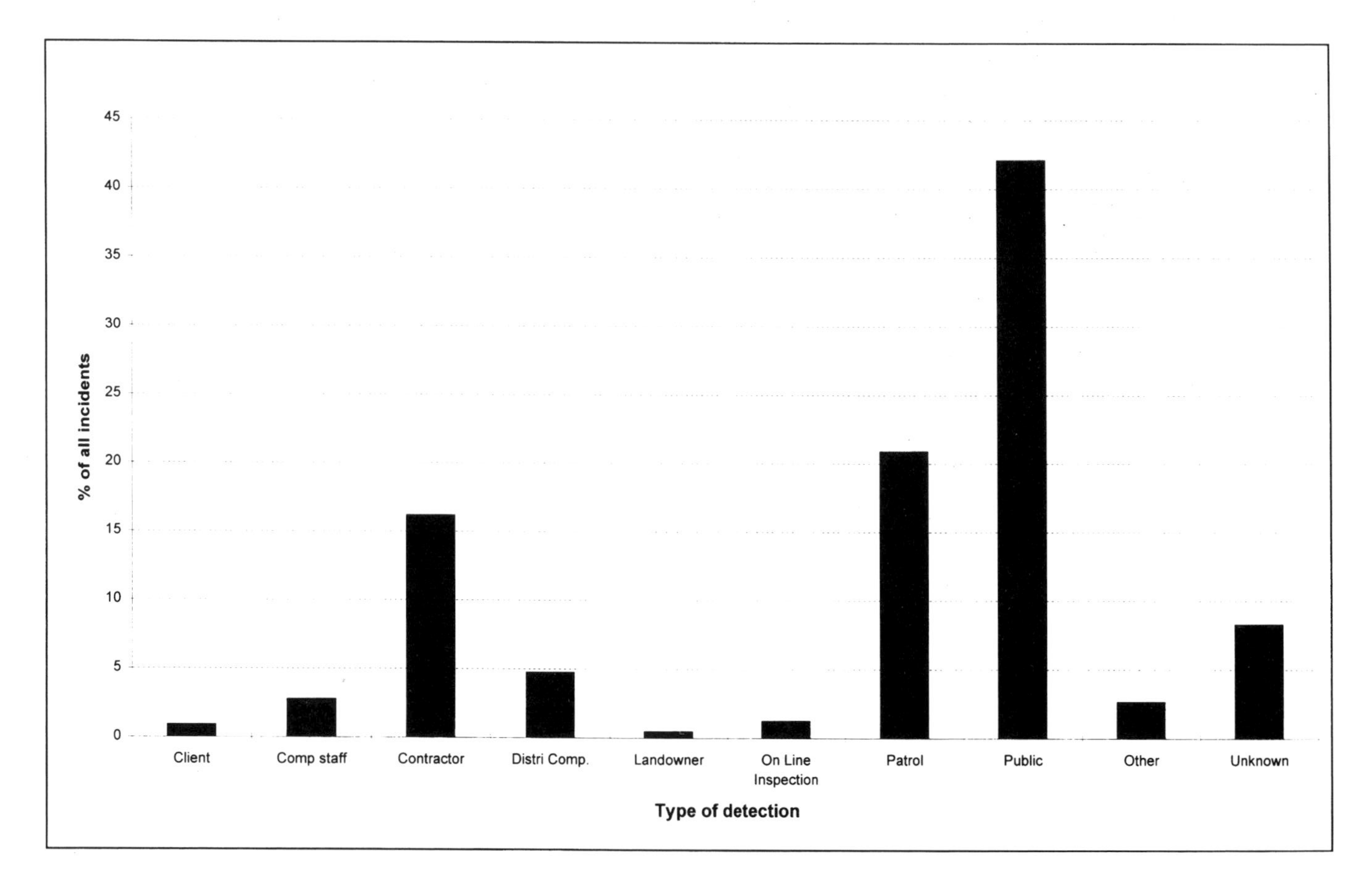

Figure 10 : Detection of incidents

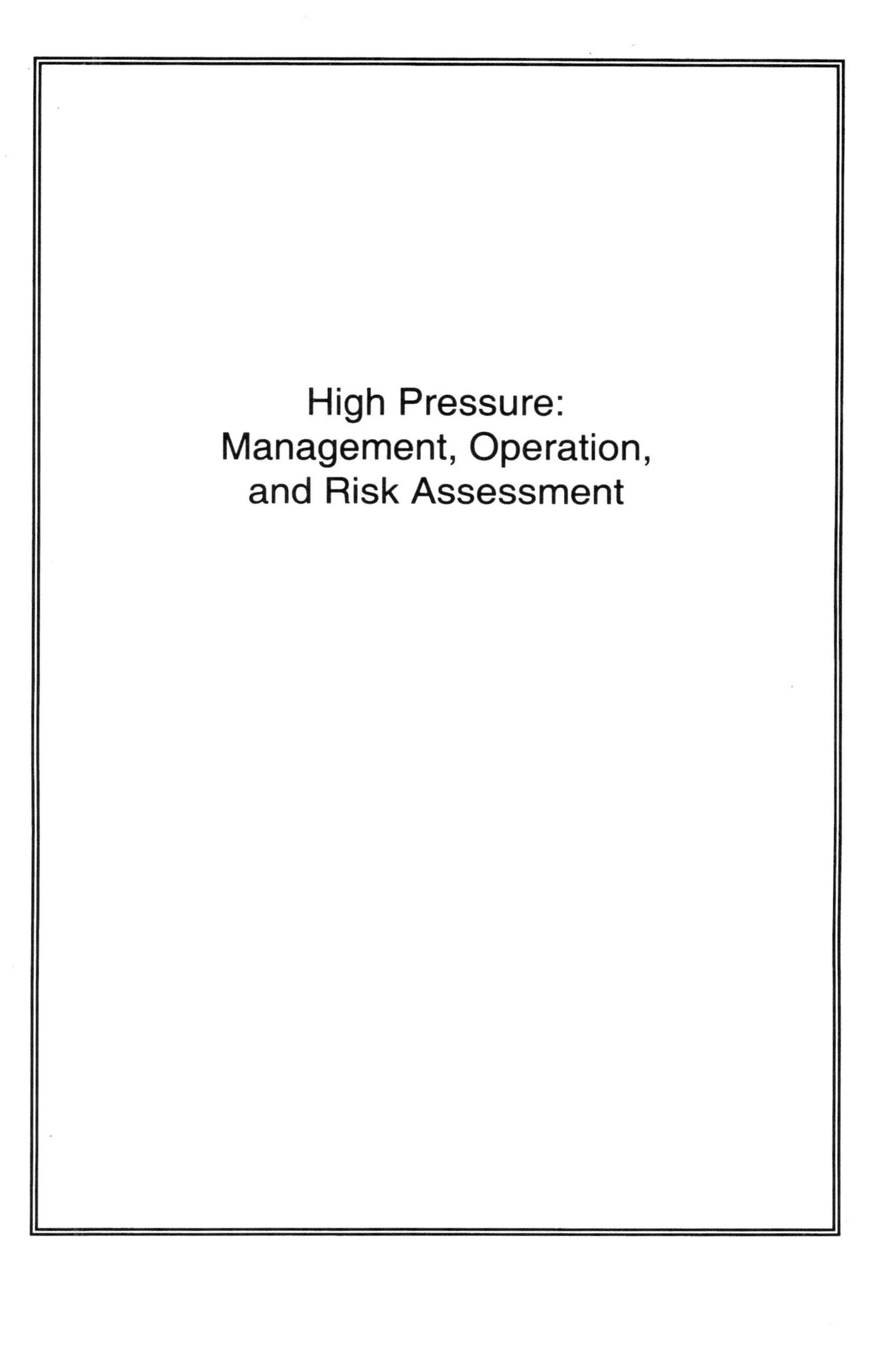

High Pressure: Management, Operation, and Risk Assessment

C571/009/99

Pipeline failure management

P A CROCKET and **R MAGUIRE**
Structural Integrity Group, ATL Consulting Limited, UK

SUMMARY

The ongoing and rapid development of pipeline systems requires an equally rapid development of new management strategies, as the performance of ageing pipelines are subject to ever greater scrutiny. Moreover, the current economic climate is one that demands a reduction in unreliability of pipeline systems due to 3rd party damage, corrosion and other mechanically induced failure mechanisms. This paper describes a Pipeline Failure Management strategy for improving the performance of ageing pipelines which identifies, evaluates and quantifies the probability of failure occurring. A methodology is also presented which can be used to assess both the probability and consequence of failure to establish a detailed risk model.

1 INTRODUCTION

Pipeline systems are continuing to evolve and expand due to the increased demand for distribution of crude and refined hydrocarbon commodity. Existing pipeline systems that were constructed in the past are now coping with new and often different operating conditions. While new pipeline systems benefit from the latest in pipeline technology and coatings, ageing pipelines are constrained by their original choice of materials and may be subject to ever greater expectations of performance as they reach and surpass their original design limitations.

Whether old or new, pipeline design is cost-optimised, which in turn means compromises must be reached to satisfy the ever-present necessity of immediate cost reduction, which in turn raises the stakes for ensuring the long term integrity of a pipeline system. The following is a good example of the fine balance that must be struck between CAPEX spend and life cycle costing to ensure safe and reliable operations.

"Corrosion occurred in a buried oil pipeline 360 km long, which had a 200 mm external diameter. A wall thickness of 6.3 mm was adequate to cope with the stress levels imposed on the pipe, but to allow for corrosion it was increased to 8.3 mm. This corrosion allowance used an extra 3000 tonnes of steel and caused a 4% loss of carrying capacity in the line" (1)

To further complicate the problem, it is unclear as to precisely what amount of wall thickness could have been accurately envisaged as corroding – the corrosion rate may have been undetermined with or without an accurate margin of error - so how far could you have cut into the margin of safety without jeopardising integrity?

This is perhaps the nub of the matter for managing the integrity of ageing pipeline systems. Pipeline designers from a generation ago did not have access to a crystal ball foretelling future events.

However, management now requires that pipeline systems are controlled with a high expectation to their current and future performance, and the expectations for ageing pipelines are greater than before. To-day's operational environment requires a rigorous and detailed Pipeline Failure Management strategy that understands and manages the risks involved and thereby provides a cost-benefit for Operators. This is a tall order but the means are now available for adapting recently developed maintenance strategies for identifying, measuring and mitigating the risk of failure and providing a measurable improvement in performance.

1.1 The Cost of Pipeline Failure

It is to be expected that lessons have been learnt from mistakes made in the past, and indeed, a review of pipeline failures shows that performance has improved over the last 25 years. The oil companies' European organisation for environment, health and safety (CONCAWE) reports that there were 6 reportable spillage incidents in 1996 on cross-country oil pipelines in Western Europe resulting in a gross oil spillage of 1414m3 (2) – roughly the equivalent of 50 petrol tankers. This may seem a large amount of leakage, but when compared against the 31,000km of network, it equates to 0,0001% of the total volume transported and perhaps seems only a trifling amount. However, when it is realised that the consequences of these spillage incidents are recorded as extreme in terms of safety as well as environmental damage, it is clear that there is an immediate requirement for an improved Pipeline Failure Management strategy.

The UK offshore oil & gas industry provides more evidence of the consequences of poorly managed pipeline systems. There have been several major failures of subsea pipelines over recent years, all of which could have been avoided. When a failure occurs on a pipeline connecting a new field then production is shutdown until the failure is identified, inspected and analysed, the cause understood, repairs completed and/or replacement pipeline installed. During this period, cash flow is halted. Operators are not keen to publish details of pipeline failures or the exact costs incurred. However, it is clear that the total bill for deferred production and repair costs caused by subsea pipeline failures can be measured in tens of millions of ECU's, and on least two occasions in the last few years, measured in hundreds of millions.

Many older platforms in the North Sea are approaching, or have reached, the end of their original design lives with production rates levelling off asymptotically. When maintenance and operation costs are trimmed, then these assets provide useful extra revenue to Operators, and have the potential to do so for the foreseeable future. The time limiting factor is the performance of the export pipeline; if it springs a leak it is unlikely that a replacement programme can be justified since typical replacement costs are in the region of 2m ECU's/km. In this situation, Operators have the potential to gain a substantial cost benefit by continuing to out-produce the original design criteria. Furthermore, abandonment costs can be put back allowing assigned revenue to be allocated elsewhere so it can be used to generate a useful 7% or 8 %.

2 FAILURE MECHANISMS

The reasons for failure can be elusive and, on the face of it, hard to predict, yet statistical examination shows that the reasons for loss of pipeline integrity fall into distinct categories. The three "big-hitters" are 3rd party damage, mechanical failure and corrosion while other causes can be grouped as operational, natural hazard and intentional damage.

2.1 3rd Party Damage

In the US (1984-87) excavation damage is thought to be responsible for 10.5% of incidents reported for onshore distribution systems. During the same period, 40.2% of damage to all pipeline distribution systems was attributed to 3rd parties' (3). CONCAWE further reported that 3rd party activity was responsible for a third of damage to pipelines (1992-96) which translated into 36% of the gross volume spilled and that the fatality recorded in 1996 was due to excavation damage of an underground gasoline pipeline which caused a large spill followed by a fire. Offshore systems are less prone to this type of failure because of the lack of heavy machinery in the vicinity, although congestion in a marine environment is acknowledged as raising the stakes of failure.

2.2 Mechanical Failure

It can also be seen from CONCAWE that the next largest failure mechanisms are mechanical failure and corrosion failure. Mechanical damage often results during, or immediately after maintenance intervention while constructional damage also falls into this category. It has been clearly documented that performance can be dramatically improved using better procedures and processes by developing a set of key performance indicators for measuring the transfer of information (4).

2.3 Corrosion

Corrosion related failures occur during the complete cycle of a pipeline's operation but predominate during later operational life and are particularly relevant to ageing pipelines. However, on closer examination it is important to understand the mechanisms that lead to corrosion failure. The nature of internal corrosion means that damage occurs in short but severe excursions; in the absence of appropriate control mechanisms, damage continues unabated until integrity is lost. Inhibition can be a critical factor. Historical evidence of corrosion failure indicates it is the random nature and amount of potential mechanisms in conjunction with the wide variety of operational conditions that provide the necessary habitat for corrosion to cause failure.

The same apparently random nature also affects external corrosion even, where static conditions prevail, whereby the metal surface reacts and corrodes at seemingly indeterminate locations. When coating breakdown and CP systems are suspected, it is important to ensure that data is gathered at regular intervals to ensure the systems are functioning properly

Pipeline failures continue to occur; at the time of writing, yet another major pipeline failure due to corrosion has been reported. Alaska's Department of Environmental Conservation has determined that a recent oil spill was caused by the corrosion of an 8" pipeline that triggered the loss of an estimated 134, 000 gallons of water/oil over a 5-day period (5).

3 STRATEGY DEVELOPMENT

Pipeline systems transport a commodity that may be pressurised, flammable and hazardous. The vast majority of post-event investigations involving a pipeline failure have shown that there has been little or no prior warning to a loss of containment, either to the cause or the consequence – therefore it is only after failure has occurred that a repair programme can be initiated. As has already been documented, a reaction, or "Fix-as Fail", strategy can be hugely costly and inefficient. A typical situation is one where the degradation mechanism has continued unabated and unobserved until a loss of containment event occurs. Under these circumstances, the only reaction management can make is "Fix-as-Fail". However, it is hard to imagine a more primitive strategy.

Management techniques in other industries have changed rapidly over recent years and it is now appropriate to adopt successful maintenance strategies where "what works well" can be applied to Pipeline Failure Management.

3.1 The P-F Interval

The P-F interval charts the condition of a component* against time (6) as shown in figure 1. It defines the interval between the occurrence of a potential failure and its decay into a functional failure. A functional failure is defined as being the state where the component cannot meet its desired standard of performance i.e. when a leak in the pipeline leads to a shutdown, or when sufficient metal is lost such that the remaining wall thickness is inadequate to retain the design pressure.

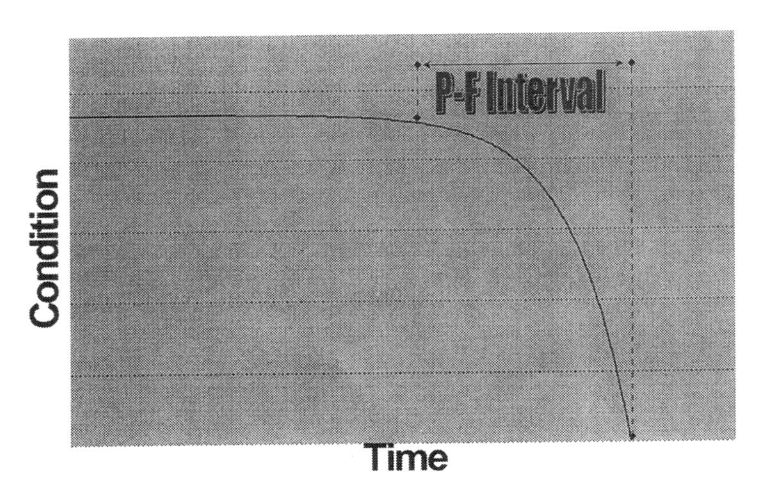

Figure 1. P-F Interval

---

* a component represents a pipeline sub-system or individual sections of pipeline where conditions are similar – applied to both external and internal conditions.

Pipelines possess a P-F interval that is dependent on the nature of the fault that requires a full understanding of the interaction between the pipeline and its operational environment. When understood and acknowledged then the ability to predict the future condition of a component is a powerful skill that is central to the development of the Pipeline Failure Management "tool-kit". What can be usefully drawn is the concept of the risk of a potential failure occurring, and the time interval available between the risk of a potential failure being identified and a functional failure occurring. When the potential risk can be identified and evaluated, then suitable action can be initiated and a cost benefit established.

3.2 The 80:20 rule

If the future condition of pipeline system components can be predicted, then where should attention focus? The answer to this question may be found by considering pareto optimality and the 80:20 rule. For pipeline systems the possibility is that 80% of problems occur in 20% of components. The next step is to identify the most critical components (the top twenty) and this can be achieved by applying risk assessment methodology

The objective of a risk assessment for ageing pipelines systems is to highlight those 20% of components that lead to 80% of problems occurring. Another way of identifying suspect components is by means of "broad brush" assessment where the objective is to screen out those 80% of components that do not pose a threat. Once potential high-risk components have been identified, a detailed assessment can be used to measure more precisely the probability of failure and calculate a time to failure. Thereby, a combination of preventive techniques and mitigating measures can be introduced to reduce the identified risk into a safe category.

The risk assessment methodology is constructed by combining probability of failure with consequence of failure. Pre-determined numerical measures that are a function of operating conditions are then applied and a matrix array completed. Statistical probabilities are used as source data where there is a clear correlation between and where enough accurate data is available. The same is true of historical data, where this is available and accurate.

The next step is to incorporate the risk assessment into a plan as detailed by the Pipeline Performance Ladder.

3.3 The Pipeline Performance Ladder

The Pipeline Performance Ladder, shown in Table 1, has been developed from the ideas, works and strategies of Edward Deming (7). It represents a structured method of five steps for moving from a "Fix-as-Fail" strategy to one that enables maximum pipeline performance as shown in figure 2.

3.2.1 Fix-as-Fail Reaction

The bottom rung of the Pipeline Performance Ladder is "Fix-as-Fail Reaction". All too often, post incident investigation reveals a situation where integrity of a pipeline was lost with no forewarning of any impending failure. This is without doubt the most costly type of incident to deal with for hydrocarbon systems as it invariably leads to a loss of flammable commodity which has a high cost impact in terms of safety, environmental protection, repair cost and deferred production. With the leak of hydrocarbon products there is the further risk of catastrophic fire and explosion.

In the offshore market, deferred production can lead to substantial losses and this cost predominates, even allowing for the environmental impact and potential loss of public support. It is perhaps surprising as to just how many failures can be categorised in this bracket, particularly when it is realised that pipeline systems are operated by multinational companies whose professed business vision is to minimise unreliability.

It should be noted that for pipeline systems that are not transporting hazardous commodity, then it may be a valid and sensible strategy to allow an item to fail before embarking on a repair/replacement programme i.e. water distribution.

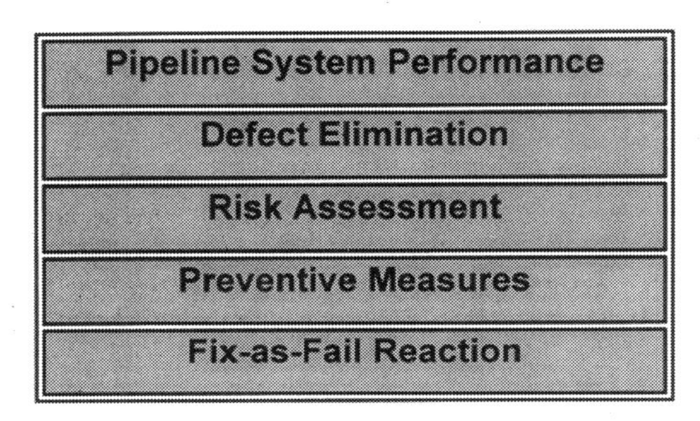

Figure 2. Pipeline Performance Ladder

3.2.2 Preventive Measures

The next step on the ladder is Preventive Measures – prevention is better than cure. Coatings and CP systems are the most obvious candidates and there have been many improvements over recent years in the materials selection, preparation, installation and covering for new pipeline systems. Improvements in distributing impressed current and the provision and spacing of sacrificial anodes on new systems are also recognised. However, the same is not necessarily true for ageing pipelines.

3.2.3 Risk Assessment

The risk assessment and predictive methodology are key to the working of the Pipeline Performance Ladder. Identification of high-risk components is a necessary step that allows resources to be allocated accordingly. It is considered in more detail in Section 4.

3.2.4 Defect Elimination

Defect Elimination is the next rung up the ladder and it shows what can be achieved by using the benefits of both preventive measures and predictive methodology. By highlighting areas of high risk it is then possible to evaluate defects before a failure has occurred and propose suitable mitigating measures to eliminate or minimise the associated risk.

3.2.5 Pipeline System Performance

The concept of Pipeline System Performance occurs equates to the position whereby a pipeline system operates at its extreme conditions all of the time with no interruptions. It carries 100% of commodity 100% of the time and is 100% reliable. In practise this does not occur, but is a necessary condition for a Pipeline Failure Management Strategy in so much that it clearly defines the direction for management to take and it is a useful condition for measuring continuous improvement over time.

It is worthwhile emphasising the point that, while the steps of this strategy development are straightforward, the application of them to Pipeline Failure Management is a new concept and fits hand in glove with the innovative ideas used for the latest developments in Performance Driven Condition Based Maintenance.

4 DEVELOPMENT OF A RISK ASSESSMENT METHODOLOGY

Predictive methodology and the application of a risk assessment are central to the successful implementation of a Pipeline Failure Management strategy. Risk assessments for pipelines are not new and can be comprehensive and consider all types and subsets of failure mechanisms (8). Their strength is the thoroughness of the technique, where all typical scenarios have a weighted numerical score factored in. The more complicated the method then greater the amount of necessary of resource to accomplish the task. When applied in a comprehensive manner, these have been proven to be of great benefit over many years. They are necessary for HAZOP's for new pipeline systems when fully comprehensive and highly detailed models can be created and applied. The downside of this approach, if there is one, is that once completed these comprehensive risk assessments cannot be easily altered to accommodate changes in operating conditions. Unforeseen predicaments require a clean sheet of paper.

It is no surprise that the benefits of this approach are equally valid when applied to ageing pipeline systems. However, for ageing pipeline systems it is possible to build relatively simple "broad brush" tool-kits that can be applied simply and used for equal effect.

4.1 Probability of Failure

Typically, the Probability of Failure matrix can be constructed using the following factors

- 3rd Party Damage
- Mechanical Damage
- External Corrosion
- Internal Corrosion
- Activity Level
- Depth & Nature Of Cover
- Population Density
- Offshore Congestion
- Buried Utilities
- Date Of Installation

Each factor has a score attached depending on the likelihood of failure occurring, which is calculated as a function of dependant conditions. There are established methods for estimating internal corrosion rates of sweet systems that consider inhibition and other factors (9). Depending on conditions, detailed subsets for atmospheric conditions, submerged pipeline, underground pipeline and induced fatigue can also be incorporated into the probability matrix, and, indeed any other factor which is perceived as being a contributing threat to the ability of the pipeline to carry out its intended function.

4.2 Consequence of Failure

The Consequence of Failure scoring can include, but not necessarily be limited to, the following factors:-

- Safety
- Environment
- Loss Of Legislative Support
- Impact On The Operator's Public Relations
- Deferred Production
- Repair Costs

Each of the consequence factors can be measured in dimensionless numbers for relative ranking. This is a powerful and effective way of establishing the critical areas or "hot-spots". Deferred production and repair costs can have monetary units attached without difficulty, but it is not so easy to attach an ECU cost to the other consequence factors.

Once compiled the results of the risk assessment are easy to understand when displayed in a graph. Figure 3 shows the results of a risk assessment. Each section, or component, of pipeline has been scored for both probability and consequence of failure on a comparative basis to calculate a risk ranking for each component. The complete set of risk ranking for all components has then been sorted in descending order to provide a clear picture of calculated risk by component identification.

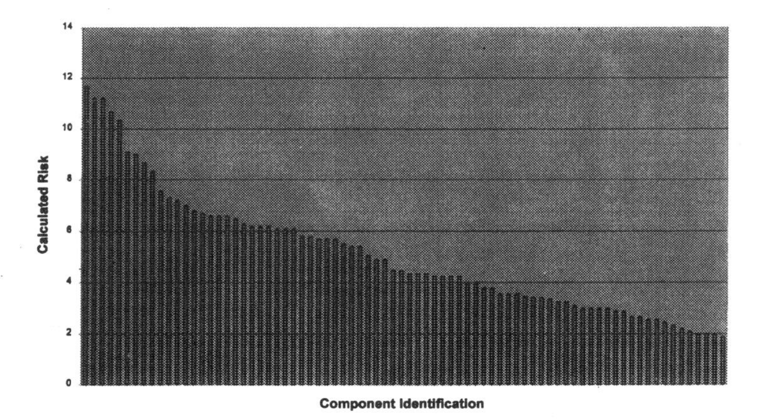

Figure 3. Risk Assessment

4.3 Derived Risk Data

The results of the risk assessment are numerical and are suitable media for analysis using spreadsheet tools. Analysis of the derived risk data is straightforward to interpret as displayed in figure 4. The derived risk data has a high intrinsic value, which is ideal for use in deciding what components would best benefit from monitoring for current and future predicted condition and for NDT surveys.

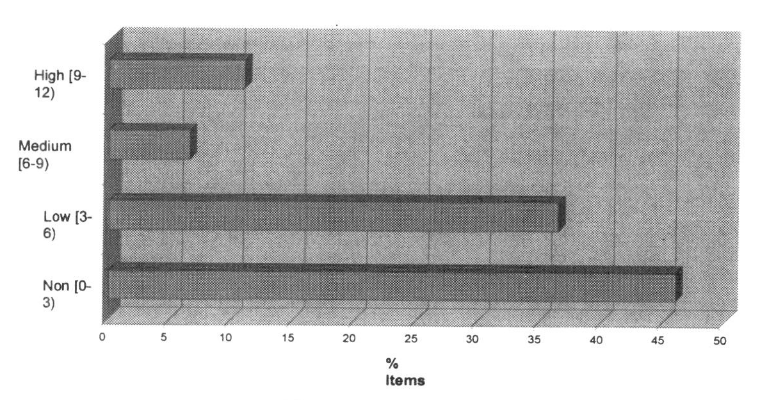

Figure 4. Derived Risk Data

5 PERFORMANCE MONITORING & THE FUTURE OUTLOOK

There is a steady development of new techniques available on the market-place for real-time corrosion monitoring of pipelines. There systems are primarily applicable to subsea pipeline systems whereby metal loss is measured and recorded to give a real-time picture of how corrosive the internal environment is. There are rapid developments in NDT techniques for pipelines that provide an insight into the condition of pipelines at any given time, and further developments whereby metal loss can be measured hourly to give an indication of current events. The results of NDT surveys can be extrapolated to indicate future conditions. Given that the cost of pipeline failure is of such magnitude, then the use of predictive methodology should be advocated as an enabling technique for achieving requisite performance.

As to the future, then perhaps Operators can borrow from the recent developments in Performance Driven Condition Based Maintenance (PD-CBM). The objective of PD-CBM is to provide an estimate of the condition of the equipment being monitored by measurement and processing of equipment parameters. Absolute performance is not relevant but rather the deterioration in condition is of far greater importance. For pipelines, the proposal is to evaluate the risk of deterioration in condition by understanding and utilising the tools of the pipeline performance ladder. The effect of this approach would be to minimise the risk of failure and avoid the enormous cost of failure.

6 CONCLUSIONS

The management of ageing pipeline systems requires a dedicated Pipeline Failure Management strategy that understands the risks involved; it is a must for to-day's Operators. The cost of pipeline failure is of such magnitude that preventive measures, predictive methodology and defect elimination should be used as enabling techniques to achieve the maximum pipeline performance. High quality data acquisition on the current and future predicted condition of pipeline systems is now possible and the interpretation of derived risk data provides a great deal of insight to pipeline condition. Performance of pipeline systems can be monitored over the long term by measuring the risk of deterioration to the pipeline components.

This approach makes good sense to an industry that is committed to reducing costs and improving efficiency. The approach is equally valid for both offshore and onshore pipeline systems. The objectives can be achieved through the development and application of a strategy that considers in detail the parameters affecting a potential failure. The tools are now available to allow quick and effective risk assessments. Consequence of failure can be established in a quick and effective manner to provide an overall risk ranking that can be evaluated and suitable mitigating methods enforced. The cost-benefits of this approach cannot be understated in terms of preventing loss and ensuring a consistent and long-term revenue stream for Operators.

7 REFERENCES

1. Trethewey & Chamberlain, Corrosion for Science & Engineering; 2nd Edition, p256, Longman, Essex, England, 1995
2. Performance of Cross-Country Oil Pipelines in Western Europe; Oil Pipelines Management Group (OP/STF-1); CONCAWE report 7/97, Brussels, 1997
3. Special Bulletin, US Department of Transportation, Research & Special Programmes Administration, Office of Pipeline Safety, 1988
4. Wheelhouse, P., Performance Indicators - How to Apply Them Successfully in Order to Evaluate Gains and Areas for Further Improvement, Preventive Maintenance, London, 1998
5. Materials Performance Volume 38 Number3, p4, NACE International, Houston, March 1999
6. Moubray, J; Reliability Centred Maintenance, 2nd Edition, p116-123, Butterworth Heinemann, Oxford, 1991
7. Deming, W.E., Out of the Crisis, MIT Press, Cambridge, Mass., 1986
8. Muhlbauer, W.K., Pipeline Risk Management Manual; 2nd Edition, p212-221, Gulf Publishing Company, Houston, 1996
9. Vuppu A.K., Jepson W.P., Study of Sweet Corrosion in Horizontal Multiphase, Carbon Steel Pipelines, 26th Offshore Technology Conference, Houston, 1994

C571/021/99

Assessing the influence of soil loading and restraint on in-service pipelines

G LEACH and **A T YOUNG**
Transco, Newcastle-upon-Tyne, UK

ABSTRACT

During its working life, a pipeline may be subjected to a range of ground loading conditions that were not anticipated within the design. Any threat to pipeline integrity must then be assessed using suitable methods and acceptance criteria.

One aspect to consider in this situation is the determination of representative soil loading and restraint values for stress analysis. This aspect is often poorly defined despite the fact that the soil loading and restraint assumptions may be found to exert a significant influence on predicted pipeline performance and integrity.

A clearer requirement for the treatment of soil loading and restraint will assist in achieving code compliant pipelines. This will avoid the inadvertent use of the code safety margin to accommodate geotechnical uncertainty and in extreme cases, pipeline failure.

The paper presents the methods of soil loading and restraint estimation in common use and identifies the soil parameters required. These parameters are then examined in terms of their range of values. The sensitivity of pipeline performance to loading and restraint bounds is illustrated through examples of uprating(increased pressure and temperature), overburden loading, and ground movement.

INTRODUCTION

In a wide range of environmental circumstances pipeline integrity can be controlled by geotechnical and geological factors[1,2,3,4,5,6,7]. These can influence the nature and form of ground disturbance and the load transfer behaviour between the pipeline structure and the soil.

New pipelines will be covered by code design considerations[8,9,10,11,12] and by safety factors to deal with uncertainties. Existing pipelines may experience disturbance which was not anticipated at their design stage.

In all cases, design methods or integrity assessment techniques are needed to assess the influence of ground movement and imposed loading on pipelines. These procedures will inevitably involve geotechnical assumptions and modelling.

European practice in buried pipe ring design is summarised in a draft European Standard prEN 1295[13]. Similar approaches are adopted in North America[14,15,16]. Techniques are generally based on Marston[17] load theory for dead load from trench and embankment conditions, Boussinesq[18] theory for live loading, and the Iowa[19] deflection equation or elastic theory for ring response. The longitudinal bending and axial response of pipelines under ground loading continues to be most often based on beam/spring structural models[20]. Shell/soil continuum[21,22] and shell/soil spring[23] modelling techniques are only occasionally utilised.

The Guidelines for the Seismic Design of Oil and Gas Pipeline Systems[24] (CGLFL) provides an important source of guidance on relating geotechnical conditions to appropriate soil spring parameters. Various studies[25,26,27,28] since this publication have produced additional useful findings. Significant work has been conducted in the Netherlands to evaluate spring parameters (e.g. [29]) which has contributed to Annex C2 of the Dutch pipeline code NEN3650[30]. This code provides a comprehensive framework and methodology for the treatment of soil loading, ground movement and soil restraint within a reliability based approach to pipeline strength and stability assessment.

A limited number of geotechnical parameters are used to define the mechanical behaviour of soils in relation to overburden loading on pipelines and soil resistance to pipeline movement. This paper considers the nature, determination and range of these parameters and attempts to illustrate their importance through a number of examples.

OVERBURDEN LOADING

The vertical earth pressure acting on a pipeline will be developed due to the weight of the overburden soil and the enhancement or reduction of loading due the interaction of the pipe ring stiffness and the supporting soil. Clarke[31] and Young and O'Reilly[32] cover the traditional approach to dealing with this problem for both rigid and flexible pipes. This is based on the experimental and theoretical work carried out by Marston[17], Spangler[19] and others at Iowa State University. Pipes in narrow trenches and flexible pipes in embankments are predicted to experience a reduced overburden pressure and rigid pipes in wide trenches or in embankments, an increased pressure.

The degree of load reduction or concentration is controlled by the assumed load transfer mechanism and an empirical factor to quantify differential pipe/soil deflection.

For pipeline construction trench widths of more than 2 to 3 times the pipe diameter[19], a soil load concentration is expected. This is produced by side fill settlement causing down drag forces on the soil prism above the pipe.

In the simplified case of differential settlement reaching the ground surface (the complete projection condition), the predictive equation for earth loading is,

$$W'_c = \frac{e^{2K\mu\frac{H}{D}} - 1}{2K\mu}\ \gamma\, D^2 \qquad (1)$$

where W'_c is the load per unit length
H is the depth of cover
D is the pipe diameter
γ is the soil unit weight
K is the ratio of horizontal soil stress to vertical soil stress (taken to be $\frac{(1-\sin\phi)}{(1+\sin\phi)}$)
μ is the coefficient of internal friction of the soil

The soil parameters required for equation 1 are γ and the angle of internal friction ϕ. These feature throughout the formulations within the Marston load theory.

In the Dutch standard NEN3650[30], this is simplified to :

$$Q_p = Q_n.(1 + 0.3\tfrac{H}{D}) \qquad (2)$$

where Q_p is the vertical soil load
Q_n is the neutral soil load (γHD)

OVERBURDEN RESTRAINT (UPLIFT)

The overburden restraint to pipe movement includes overcoming both the gravitational action of the soil and pipe weight and the resistance in mobilising the strength of the overburden soil.

A wide range of formulations have been published for the prediction of surface breakout resistance to a buried object (e.g. anchor plates, cylinders). These are based on either vertical failure planes[17,19,30,33,34], inclined failure planes[35,36], curved failure planes[37] or cavity expansion theory[30,38].

Recommendations from the CGLFL guidelines[24] for ultimate uplift resistance q_u (force per unit length) are,

$$q_u = \bar{\gamma}\, Z\, N_{qv}\, D \text{ for sands} \qquad (3)$$
$$q_u = s_u\, N_{cv}\, D \text{ for clays} \qquad (4)$$

where Z is the depth to pipe axis
$\bar{\gamma}$ is the soil effective unit weight
s_u is the undrained shear strength of the soil

N_{qv} & N_{cv} are uplift factors dependent on the burial geometry and the angle of internal friction ϕ of the soil.

Alternative guidance in NEN3650[30] recommends the use of the Marston load theory for shallow pipelines and a cavity expansion formulation for deep pipelines.

The soil parameters of interest are the unit weight (total - γ or submerged - γ'), the angle of internal friction (drained - ϕ' or undrained - ϕ_u) and the soil cohesion (drained - c' or undrained - c_u or s_u).

LATERAL RESTRAINT

The ultimate bearing capacity of a soil subjected to lateral pipeline displacement[24,30] is typically predicted using a solution published by Brinch Hansen[39] for horizontally displaced piles.

The equations for maximum horizontal earth pressure are of exactly the same form as equations 3 and 4. The required soil parameters are also the same.

In the case of pipelines, the lateral restraint may be influenced by both the local backfill alongside the pipe and the adjacent natural ground properties. Work by Ng[40] suggests that in cohesive soil the influence of the natural ground should be considered when the trench width is less than approximately three pipe diameters.

DOWNWARD RESTRAINT

The CGLFL guidelines[24] suggest the use of the conventional bearing capacity calculation[41] for a strip footing,

$$q_u = cN_cD + \bar{\gamma}ZN_qD + \tfrac{1}{2}\gamma D^2N_\gamma \qquad (5)$$

where q_u is the ultimate force per unit length
c is the soil cohesion
N_c, N_q & N_γ are factors depending on the soil angle of internal friction (sourced from Meyerhof[42])

An alternative formula based on an extended solution attributed to Brinch Hansen[43] is recommended by NEN3650[30].

AXIAL RESTRAINT

The preferred expression for the axial force per unit length of pipe due to the sliding friction of the backfill on the pipe coating is taken from NEN3650[30],

$$W = \pi D\,((\tfrac{1+K}{2}).\sigma_k.\tan\delta + a) + (Q_{eg} + Q_{vul} + Q_{op}).\tan\delta \qquad (6)$$

where W is the ultimate sliding friction force per unit length
σ_k is the vertical intergranular soil stress at pipe axis level
δ is the angle of sliding friction of the soil backfill on the pipe coating
a is the adhesion of the soil backfill on the pipe coating
Q_{eg} is the load per unit length due to the pipe weight
Q_{vul} is the load per unit length due to the pipe contents
Q_{op} is the load per unit length due to buoyancy

The required soil parameters are the unit weight γ (controlling σ_k), the angle of internal soil friction ϕ' (controlling K) and the adhesion a and angle of sliding friction δ of the backfill on the pipe surface.

STIFFNESS RESPONSE

The stiffness response of the ground to pipe movement is typically represented by a bi-linear model. This is convenient for beam/spring piping stress analysis and recognises that more elaborate response simulation is beyond current justification.

Uplift

The CGLFL guidelines[24] refer to the experimental findings of Trautmann & O'Rourke[44]. A hyperbolic force/displacement relationship is recommended from experimental tests in dry sand. The equation is expressed as follows,

$$q = \frac{z}{A'' + B''z} \qquad (7)$$

where q is the uplift resistance
z is the uplift displacement
$A'' = 0.07\frac{z_u}{q_u}$
z_u is the displacement at ultimate resistance
q_u is the ultimate uplift resistance
$B'' = \frac{0.93}{q_u}$

A bi-linear approximation to equation 7 can be obtained by a linear response fitted through the $\frac{q_u}{2}$ point on the curve. Without derivation the initial linear stiffness is $7.64\frac{q_u}{z_u}$. A second stiffness of negligible magnitude is adopted when the ultimate resistance is reached. The CGLFL guidelines[24] suggest that z_u may lie in the range 0.01-0.02Z for granular soils and 0.1-0.2Z for cohesive soils.

An alternative approach for uplift is to assume a near rigid stiffness response followed by a cut-off at the ultimate resistance.

Lateral
The CGLFL guidelines[24] use a force/displacement relationship of the same form as equation 7,

$$p = \frac{y}{A' + B'y} \qquad (8)$$

where p is the lateral resistance
y is horizontal pipe displacement
$A' = 0.15\frac{y_u}{p_u}$
$B' = \frac{0.85}{p_u}$
p_u is the ultimate resistance
y_u is the displacement to reach p_u

The displacement y_u is expected to lie in the range 0.02-0.10Z for granular soils and 0.03-0.05Z for cohesive soils.

The NEN3650[30] code suggests that as a preliminary estimate the linear lateral stiffness response can be taken to be 0.7 times the downward stiffness.

Downward
The CGLFL[24] guidelines suggest that the ultimate resistance will be developed at a downwards displacement of 0.1-0.15D. NEN3650[30] suggests a vertical bedding stiffness $k_v = \frac{2p_{we}}{0.07D}$ for peat and undrained response in clay, and $k_v = \frac{p_{we}}{0.07D}$ for sand and drained behaviour in clay. p_{we} is the bearing failure pressure.

Axial
The maximum sliding resistance is expected[24,30] to occur at 1-5mm in sand, 2-10mm in clay and 4-18mm in peat.

GEOTECHNICAL PARAMETERS

A number of geotechnical parameters are required for the prediction of earth loading and soil restraint on pipelines. Some soil parameters (density and angle of internal friction) appear in all of the conditions, whilst others (e.g. soil cohesion) appear in most situations. The engineer must select appropriate values for the parameters for the specific pipeline or installation under study.

Density
Bulk density observations on cohesive pipe backfill materials in the UK are presented in figure 1. Cohesive materials are found to be dominant in the drift deposits of the UK and account for approximately 80% of backfill. The estimated 95% confidence limits for cohesive backfill density provide the range 1780-2230 kg/m^3. Within this range, analysis suggests that density is linked to water content, plasticity and the proportion of coarse particles in the material grading.

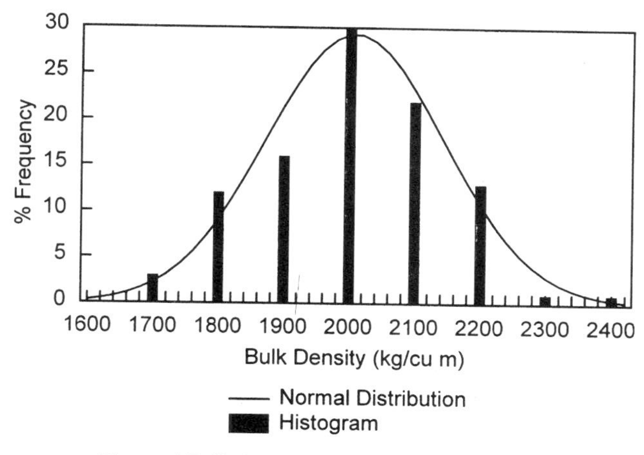

Figure 1 Bulk density distribution in cohesive backfill

Angle of internal friction

The strength parameter for a cohesionless soil is the angle of internal friction. This depends on the soil mineralogy, the angularity of the particles, the particle size distribution (grading), the packing density and the stress level. The mineral interparticle sliding friction is typically taken to be 26° for quartz and 37° for feldspar[45]. The minimum angle of internal friction ϕ_{ult} (ultimate or critical state) represents the combined influence of the mineral friction and a contribution due to particle interlock in a loose packing state. This latter contribution has been quoted[46] as approximately 5° to 6° which agrees closely with ultimate friction angles of 33° for quartzitic sand and 40° for felspathic sand quoted by Bolton[47]. Uniform sands with rounded grains and silty or fine sands may exhibit an angle of internal friction as low as 26° to 27° [48,49].

The relative density, the material confinement during shear, and the intergranular stress at failure will combine to dictate the peak angle of internal friction ϕ_{max} for a granular soil. For sands, Bolton[47] offers the expression,

$(\phi_{max} - \phi_{ult}) = 5.I_R$ *for plane strain or* $3.I_R$ *for triaxial conditions*

where $I_R = I_D(10 - \ln p') - 1$

I_D is the relative density (0-1)

p' is the effective stress at failure (kN/m²)

Bolton[47] also suggests that the maximum value of $(\phi_{max} - \phi_{ult})$ should be limited to 20° for plane strain and 12° for triaxial conditions. This would lead to a maximum plane strain angle of internal friction of 53° for a quartzitic sand and 60° for a feldspathic sand. In the context of the safe bearing capacity of structures these values would be rarely utilised due to concern over the influence of progressive failure in very dense materials. In contrast, for an upper bound soil restraint estimation for buried piping, the direct application of these values would be acceptable.

For the purpose of direct measurement, the direct shear test offers the most convenient and cost effective testing solution. The angle of internal friction from this test will typically produce a value 0° to3° higher than in triaxial conditions[45,49] and 2° to 9° lower than in plane strain conditions[50].

The angle of internal friction for a cohesive soil may be defined in two ways depending on whether soil restraint is predicted in terms of total stress or effective stress (total stress - porewater pressure). For short term, undrained soil response, a total stress angle of internal friction ϕ_u may be linked with the undrained cohesion c_u to define the material strength. The appearance of an undrained angle of internal friction tends to be associated with unsaturated deposits of marginal classification. This is illustrated in figure 2 by the distribution of measured undrained friction angles from strength tests on Scottish Boulder Clays.

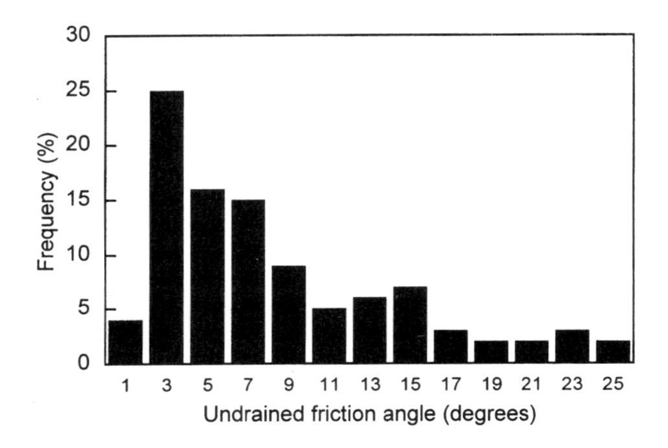

Figure 2 - Undrained friction values for Boulder Clays in Scotland

In relation to soil restraint calculations for pipelines, the drained or effective angle of internal friction for a cohesive soil is most often required to allow an estimation of the horizontal stress (or pressure) in the soil. This then allows the average soil contact pressure on the pipe coating to be assessed for the calculation of axial sliding resistance. The peak friction value for cohesive soil typically varies with clay mineralogy, clay fraction, stress level and stress history. Values vary approximately from 34° ± 6° for low plasticity soils to 24° ± 6° for high plasticity soils.

Undrained shear strength

Consideration of undrained shear strength is typically restricted to cohesive soils and is generally defined in terms of an undrained cohesion c_u and an undrained friction angle ϕ_u. For saturated or near saturated cohesive soils the value of ϕ_u is usually close to or equal to zero and is rarely quoted if less than 5°. For most cohesive soils in the UK it is adequate for the undrained shear strength to be taken as c_u.

The undrained shear strength of both cohesive natural ground and cohesive backfill (in the piping construction trench) is of interest for the estimation of ground restraint. Natural ground strength varies greatly although some moderation of range can be expected due to the placement of buried piping within the weathered zone of most deposits.

Cohesive backfill is generally weaker than the original natural material. This is due to the action of remolding on the soil strength (referred to as sensitivity) and poor material compaction.

Data on a low sensitivity glacial deposit, *Boulder Clay*, illustrates the contrast in backfill and natural ground condition and strength. Figure 3 shows the recorded distribution of shear strength on 110 samples of boulder clay taken from shallow depth in Scotland. The mean shear strength is 71 kN/m^2.

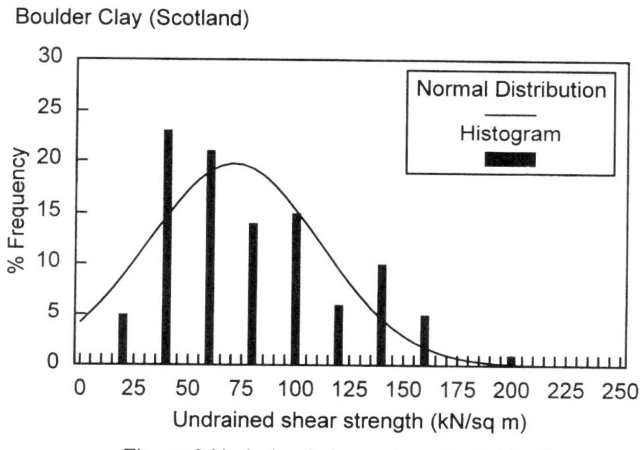

Figure 3 Undrained shear strength distribution

The corresponding distribution for the undrained shear strength of cohesive backfill of boulder clay origin is presented in figure 4. The average recorded shear strength is 17 kN/m^2 from 130 measurements.

The explanation for the consistently lower strength in the backfill must be due to a factor other than remolding, since boulder clays are generally recognised as exhibiting a sensitivity close to unity.

A comparison of the shear strength and moisture content of natural and backfill soils of boulder clay origin is presented in figure 5. This indicates that the moisture content of trench backfill is consistently higher than that of the parent material in its natural condition. This increased moisture content is attributed to water ingress into poorly compacted backfill and accounts for the reduced strength relative to the natural material.

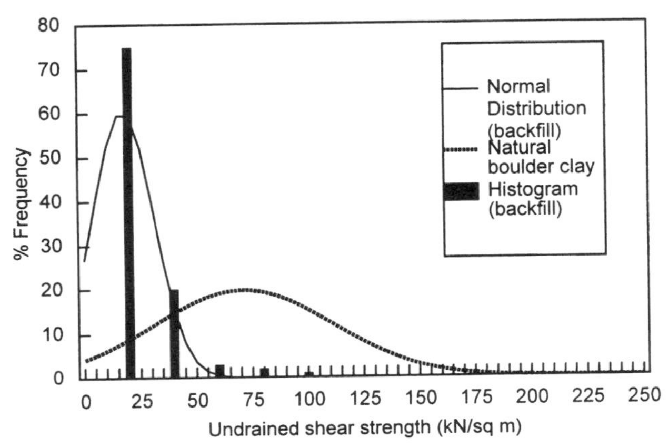

Figure 4 Undrained shear strength distribution in boulder clay backfill

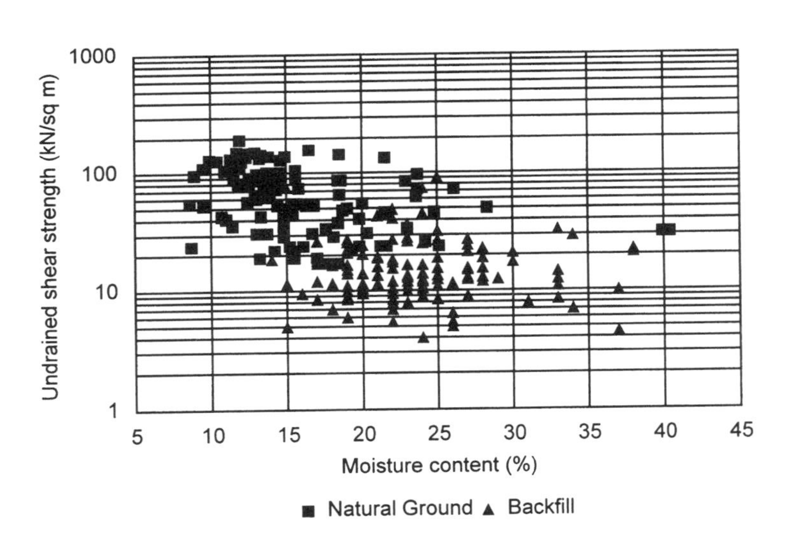

Figure 5 Undrained shear strength/moisture content link for boulder clays

Soil/pipe adhesion and sliding friction

The backfill type and condition and the pipe coating will influence the sliding resistance at the soil/pipe interface.

In the case of natural granular soils with negligible fines, or washed imported aggregates, the absence of soil adhesion can normally be assumed. For cohesive soils, the sliding resistance parameters are found to be dependent on the moisture content, plasticity and coating type.

Indicative values from direct shear tests on common coating types are presented in Table 1.

Table 1 - Adhesion & friction angles

Backfill	Adhesion (kN/m^2)	Peak angle of sliding friction (degrees)
Granular	0 - 6	24 - 47
Cohesive	0 - 18	12 - 38.5

The shear strength of the soil-pipe interface can adopt a wide range of values and coating type exerts a strong influence. Coal tar enamel exhibits the highest adhesion and angle of friction, polyethylene displays the lowest values and fusion bonded epoxy exhibits an intermediate response.

Pipeline trench width
The extent of backfill on each side of a pipe and its condition relative to the adjacent natural or fill material will influence the magnitude of the overburden loading on the pipe and the restraint offered to lateral pipe movement. In urban areas the excavation is generally kept to a minimum and support is usually provided to vertical trench walls. In rural areas the trench is often of a 'V' shaped cross section or is benched to aid stability.

Watertable
Water seepage in open excavations or water strikes in exploratory boreholes may provide an indication of current water levels. For seasonal extremes a long term monitoring scheme is required using standpipes or piezometer installations.

In the absence of evidence on groundwater depth variation, the ground restraint calculation process will normally proceed on the basis of extreme assumptions. For minimum ground resistance the watertable would be assumed to be at the ground surface and for maximum restraint a water depth of at least two pipe diameters below pipe invert level would be used.

PARAMETER SELECTION AND UNCERTAINTY

Quality of ground investigation
The volume and accuracy of soil property information for a site will influence the range of calculated soil loading and restraint values. For situations where it is considered essential to work with reliable values, a comprehensive ground investigation will be required. In the case of existing installations this will include a study of the nature, condition and extent of backfill associated with the original construction.

For new installations an assessment will be required of the suitability of the natural ground for re-use as a backfill material. The alternative option of an imported a granular material as pipe bedding and surround material will also need to be considered. The chosen material acceptance criteria and the defined material placement requirements will influence the uncertainty over the final backfill condition and hence the range of basic soil parameters chosen for piping stress analysis.

Thorough ground investigation and testing may allow mean parameters to be adopted for the ground restraint determination. Alternatively, a relatively narrow range of restraint bounds will be identified from the site data variability.

For situations where excavation and backfilling requirements are undefined or where an existing installation is not investigated, the piping stress analysis will need to proceed with a greater level of uncertainty over the loading or restraining influence of the soil.

Construction stress

Geotechnical conditions are one factor contributing to construction stresses within piping and pipelines. Uneven trench bottom conditions and the loading influence of the soil overburden can lead to the development of longitudinal bending stresses. The Dutch NEN3650[30] code offers guidance on the level of construction settlement that might be expected depending on the ground conditions. Alternatively, direct measurements of pipeline stress level will provide an indication of construction stress.

Confidence level & contingency factors

The dilemma with soil property selection is whether to use conservatively low or high values, mean values, or assessed values based on interpretation and judgement of site data. The latter option is essentially a mean value approach using a reduced, more representative data set for the site.

In the mean or assessed value approach the uncertainty in relation to the predicted pipeline performance differs according to the method of integrity assessment. In a permissible stress approach the uncertainty must be absorbed with the overall '*lumped*' factor safety. In a reliability based approach the uncertainty is handled by contingency factors applied to the calculated soil loads and restraints. These are applied to increase loads (overburden and settlement) and to increase or reduce restraints depending on the loading regime and pipe layout.

The use of conservatively low or high values can be based on visually setting bounds to data sets or by attempting to place bounds at a declared confidence level. The conservatism is associated with an implicit assumption that the soil properties are uniform at the selected level whereas in fact the data will form a distribution.

The advantage of working with conservative bounds is that the '*lumped*' factor safety is no longer used to embrace the uncertainty over geotechnical factors. This is the preferred approach since there is generally a lack of clarity in most codes on whether or not a degree of geotechnical uncertainty can be accommodated within the design factors.

An example of mean and bounding value selection from in-situ testing results on a sand deposit is presented in figure 6.

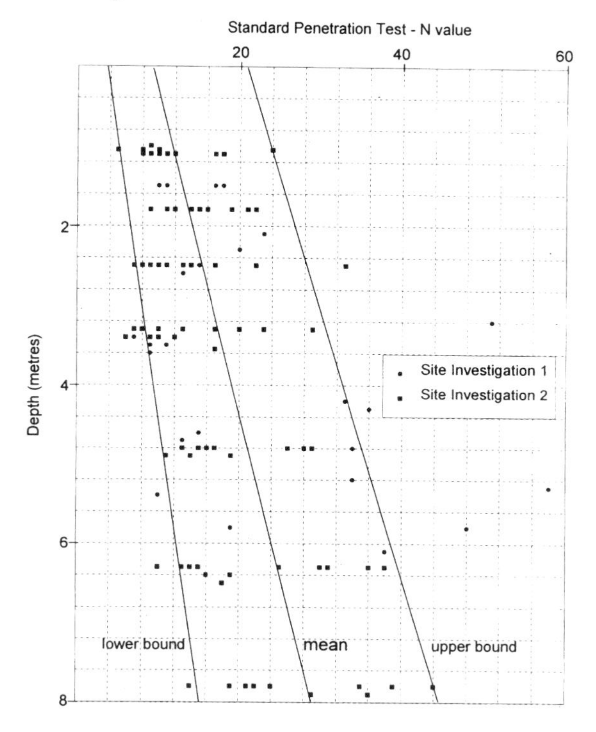

Figure 6 Example of 95% confidence bounds on field data

SENSITIVITY OF PIPING RESPONSE

Three examples are presented to illustrate the influence of soil properties on pipeline behaviour.

Internal pressure and temperature

High pressure transmission pipelines frequently experience significant pressure and thermal loads especially when located downstream of a compressor discharge. Pressure and thermal strains develop in the pipeline due to the presence of bends, which cause a change in restraint on the pipeline through a change in pipeline direction. In this situation, it is the restraint offered to the pipeline by the adjacent soil mass which controls the pipeline behaviour.

Uprating projects on high pressure gas transmission pipelines have recently required a study of the integrity of bends close to compressor discharges. A maximum pressure of 85 bar and maximum temperature of 50° C were considered to apply to 36" diameter pipelines in the study area.

The principal area of consideration covered pipelines in Scotland. From a range of soil property values measured on glacial tills from pipeline construction projects in Scotland, lower and upper bound soil restraint values were calculated at a 95% confidence level.

Figure 7 shows the influence of the soil restraint on the predicted stress level (presented as the percentage of acceptable stress). The pipeline geometry was considered to be comprised of two side bends separated by a distance of 10 metres. This is typical of a route alignment configuration used at road, rail or river crossings.

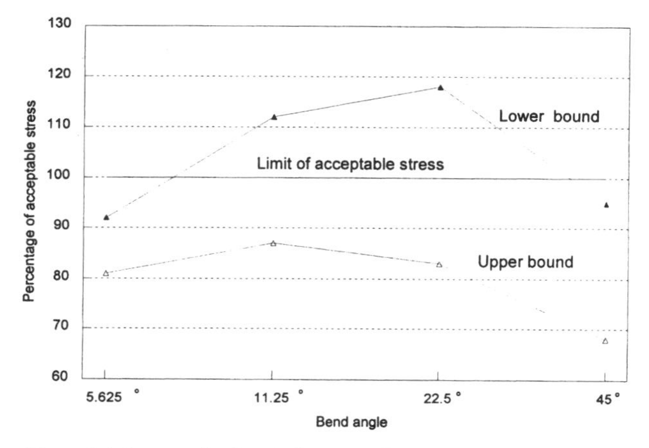

Figure 7 Influence of soil restraint on predicted stress at bends subjected to internal pressure and temperature loading

It is clear from the figure that the predicted stress level remains within the acceptability limit for the upper bound restraint condition. A significant increase in stress is predicted for soil restraint set at the lower bound level and the stress exceeds the permitted level for bend angles in the range 11.25° and 22.5°. This leads to the requirement of a more detailed assessment for some installations. This might include site specific soil property determination, evaluation of temperature decay profiles from the compressor discharge, or the consideration of ambient temperature at the time of the pipeline installation.

Overburden loading

The standard depth of cover on pipelines is typically between 0.9 and 1.2 metres to reduce the incidence of damage from third party interference. External loads, which are additional to the standard design condition, might arise from installation at deeper cover, from increased cover due to embankment construction, or from surface surcharge loading due to vehicles, water, spoil or construction materials.

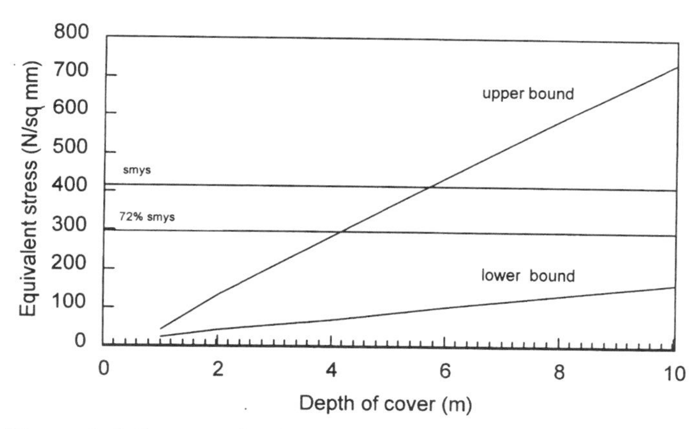

Figure 8 Influence of soil property bounds on cross-section stress

Under the action of non-standard or abnormal external loads, both the longitudinal and cross-sectional response of the pipeline needs to be considered. The properties of the soil surrounding the pipeline will control the transmission of load onto the pipeline and the pipeline response.

In figure 8, the maximum equivalent stress developed in a 36"diameter pipe ring due to soil loading on the pipe cross-section is plotted against a varying depth of cover for both lower and upper bound soil properties relevant to Coal Measures drift deposits. This illustrates that the contribution from soil loading to the pipeline stress state can be significant and that the predicted response is sensitive to soil parameter assumptions. As a consequence of the possible importance of this type of loading, some codes[30,51] require the influence of soil overburden to be specifically considered when the cover depth exceeds 2.5-3.0 metres.

Measures to reduce the development of stresses from overburden loading may include pipeline installation in a narrow trench, high compaction of the side fill, selective use of a slab or the use of a low density backfill.

Ground movement

Pipelines can be affected by a wide variety of ground movement events. Those associated with underground excavation and slope instability tend to be of primary interest. The ground movement zone produces the soil loading that induces the pipeline displacement. This is resisted by the pipeline structure and by restraint from the soil adjacent to the area of instability.

The effect of longwall coal mining subsidence on a 48" pipeline in South Wales has been directly measured using strain gauges at six locations. In addition the ground settlement profile was also recorded. This location was of particular interest due to the presence of a geological fault which was expected to distort the typical ground settlement profile produced by longwall mining.

The measured ground subsidence was applied to a numerical model of the pipeline. The response of the pipeline to the applied subsidence was calculated for upper bound and lower bound soil properties relevant to Coal Measures drift deposits. The calculated bending stress profiles are illustrated in figure 9 and compared to the measured strain gauge readings converted to bending stresses. It can be seen that the lower bound soil properties lead to an under prediction of stress compared to the measured values. The predicted stress profile based on upper bound soil properties illustrates a close fit to the data. The actual pipeline was laid in lytag granular fill on weathered bedrock and this condition is considered to be very close to the upper bound case.

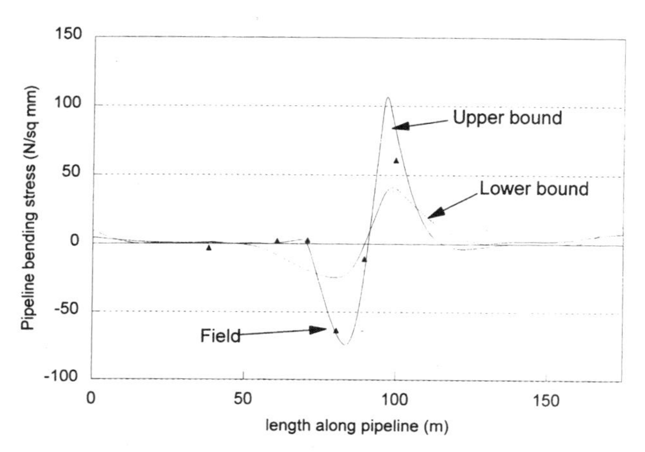

Figure 9 Comparison of predicted bending stress with field data due to mining subsidence

The selection of soil restraint is observed to significantly influence the magnitude of the calculated pipeline stress. This could have implications for decisions on monitoring, protection or diversion requirements.

In a predictive situation the calculated range of pipeline stress would be further extended by the uncertainty over the magnitude and form of ground movement.

DISCUSSION

The example problems of internal pressure and temperature loading, overburden loading and ground settlement loading illustrate that the stress level in a pipeline can be sensitive to geotechnical conditions. Most pipeline codes expect geotechnical aspects to be handled on the basis of good engineering practice. This will inevitably lead to some inconsistency in the reliability of predicted soil loading, settlement or ground resistance. This could lead to pipelines out of code or in exceptional circumstances to pipeline failures or to an unserviceab'e deformation condition.

The pipelir.e codes that offer a reliability based integrity approach are generally specific about the requi:ements for achieving a stated reliability level for soil loading, settlement and restraint This reduces the potential for inconsistency.

In the application of traditional permissible stress codes, there is also a good case for geotechnical aspects to be selected according to a target confidence level and for appropriate consideration to be given to the sensitivity of pipeline integrity at both lower bound and upper bound conditions. This would eliminate the possibility of the inadvertent use of the code safety factor to accommodate geotechnical uncertainty.

CONCLUSIONS

A range of algorithms and calculation methods for the prediction of soil loading and ground restraint on pipelines are available. These are considered to represent good engineering practice in relation to geotechnical aspects of pipeline integrity. However, the selection of soil parameters for these approaches is dependent on the data available and on the judgement and experience of the engineer. This inevitably leads to inconsistency in the quality and reliability of pipeline integrity assessments involving geotechnical factors.

Reliability based methods of pipeline integrity assessment are seen to place the geotechnical aspects within a more clearly defined framework and potentially achieve their target of a known pipeline safety level. This type of reliability approach to geotechnical aspects can still be applied within the traditional permissible stress based pipeline codes. However, the potential sensitivity of pipeline response to geotechnical assumptions would justify a more prescriptive requirement in the codes. This would ensure code compliant pipelines and should further ease the inevitable transition to the full adoption of reliability based design.

REFERENCES

1. Venzi S, Malacarne C & Cuscunà S. *Development of an expert system to manage the safety of pipeline in unstable slopes.* Proc. Int. Conf. on Offshore Mechanics and Arctic Engineering. Volume 5. Pipeline Technology. ASME 1993.

2. Trigg A. & Rizkalla M. *Development and Application of a Closed Form Technique for the Preliminary Assessment of Pipeline Integrity in Unstable Slopes.* Proc. 13th Int. Conf. on Offshore Mechanics and Arctic Engineering. ASME. 1994.

3. Herbert R. & Leach G. *Damage Control Procedures for Distribution Mains.* 56th Autumn Meeting of the Institution of Gas Engineers. London. November 1990.

4. Fearnehough G. & Middleton E. *Pipelines take the strain.* R&D Digest No. 11. Britsh Gas plc. 1992.

5. *The Palaceknowe Pipeline Diversion Failure.* Information Sheet. British Gas R&T. 1995.

6. Iwata T. *Current Aseismatic Design Techniques on Buried Pipelines. Presentation of Conventional Standards and their Operational Comparisons.* Tokyo Gas Company Ltd. October 1990.

7. Middleton E. & Henderson M.J. *A Tale of Two Faults*. Gas Association of Wales. October 1994.

8. ASME B31.8-1992 Edition. *Gas Transmission and Distribution Piping Systems.* The American Society of Mechanical Engineers, 345 East 47th Street, New York, N.Y.10017.

9. BS8010:Section 2.8:1992. *Code of Practice for Pipelines. Part 2. Pipelines on land: design, construction and installation. Section 2.8 Steel for oil and gas.* British Standards Institution. 1992.

10. prEN 1594. *Pipelines for gas tranmission.* European Committee for Standardisation. 1994.

11. IGE/TD/1 Edition 3: 1993. *Recommendations on Transmission and Distribution Practice.* Communication 1530. The Institution of Gas Engineers. London. 1993.

12. *The Recommended Standards for Earthquake Resistant Design of Gas Pipelines.* Japan Gas Association. March 1982.

13. prEN 1295. *Structural design of buried pipelines under various conditions of loading.* Draft European Standard. 1994.

14. Moser A. P. *Buried Pipe Design.* McGraw-Hill. 1990.

15. American Concrete Pipe Association. *Concrete Pipe Design Manual.* 1992.

16. American Water Works Association. *Steel Pipe-A Guide for Design and Installation (M11).* 1989.

17. Marston A. *The theory of external loads on closed conduits in the light of the latest experiments.* Proc. Highway Research Record. Vol. 9. 1930.

18. Boussinesq V. J. *Application des potentiels à l'étude de l'équilibre, et du mouvement des solides élastiques avec des notes etendues sur divers points de physique mathématique et d'analyse*. Paris. 1885. (Gauthier-Villars).

19. Spangler M.G. *Underground Conduits-An Appraisal of Modern Research.* Proc. ASCE. Vol. 73. 1947.

20. *PLE-micro-CAD.* Expert Design Systems bv. 1987.

21. Ingraffea A.R., O'Rourke T.D. & Stewart H.E. *Technical Summary and Databases for Guidelines for Pipelines Crossing Railroads and Highways.* Report No. GRI-91/0285. Gas Research Institute. December 1991.

22. Bolzoni G., Cuscunà S. & Perego U. *Physical and mathematical modelling of pipeline behaviour in landslide areas.* 9th PRC/EPRG Biennial Joint Technical Meeting on Line Pipe Research. 11-14th May, Houston, Texas. 1993.

23. Yang K. *Shell model FEM analysis of buried pipelines under seismic loading.* Bulletin of Disaster Prevention Research Inst. Vol. 38. Part 3. No. 336. 1988.

24. *Guidelines for the Seismic Design of Oil and Gas Pipeline Systems.* Committee on Gas and Liquid Fuel Lifelines. ASCE. 1984.

25. Ng P.C.F., Pyrah I.C. & Anderson W.F. *Prediction of soil restraint to a buried pipeline using interface elements.* Proc. 6th International Symposium on Numerical Models in Geomechanics - NUMOG VI. Montreal. Canada. July 1997.

26. Dickin E.A. *Uplift Resistance of Buried Pipelines in Sand.* Soils and Foundations. Vol. 34. No.2. Japanese Society of Soil Mechanics and Foundation Engineering. June 1994.

27. Rizkalla M., Trigg A. & Simmonds G. *Recent Advances in the Modelling of Longitudinal Pipeline/Soil Interaction for Cohesive Soils.* Proc.15th International Conference on Offshore Mechanics and Arctic Engineering. Volume V. Pipeline Technology. ASME 1996.

28. Cappelletto A., Tagliaferri R., Giurlani G., Andrei G., Furlani G. & Scarpelli G. *Field Full Scale Tests on Longitudinal Pipeline-Soil Interaction.* International Pipeline Conference. Volume II. ASME 1998.

29. *Grondonderzoek Gedrag Van Buisleidingen In Klei. Onderzoek Uitgevoerd Te Kesteren In 1984.* Laboratorium Voor Grondmechanica. Delft. January 1985.

30. NEN3650. *Eisen voor stalen transportleidingsystemen* (Requirements for steel pipeline transportation systems). Nederlands Normalisatie-instituut. Delft. 1992.

31. Clarke N.W.B. *Buried Pipelines. A manual of structural design and installation.*Maclaren and Sons. London. 1968.

32. Young O.C, and O'Reilly M.P. *A guide to design loadings for buried rigid pipes.* HMSO. 1987.

33. Meyerhof G.G. & Adams J.I. *The Ultimate Uplift Capacity of Foundations.* Canadian Geotechnical Journal. Vol. 5. No. 4. 1968.

34. Ladanyi B. & Hoyaux B. A Study of the Trap-Door Problem in a Granular Mass. Canadian Geotechnical Journal. Vol. 6. No. 1. 1969.

35. Murray E.J. & Geddes J.D. *Uplift of Anchor Plates in Sand.* Journal of Geotechnical Engineering. ASCE. Vol. 113. No. 3. 1987.

36. Saran Swami, Ranjan Gopal & Nene A.S. *Soil Anchors and Constitutive Laws.* Journal of Geotechnical Engineering. ASCE. Vol 112. No. 12. 1986.

37. Shimamura K. & Takagi N. *Vertical earth pressure on buried pipes in a projection condition.* Proc. 13th Int. Conf. Offshore Mechanics & Arctic Engineering. Vol. 5. Pipeline Technology. ASME 1994.

38. Vesic A.S. *Breakout Resistance of Objects Embedded in Ocean Bottom.* Journal of the Soil Mechanics and Foundations Division. ASCE. Vol. 97. No. SM9. 1971.

39. Brinch Hansen J. *The Ultimate Resistance of Rigid Piles Against Transversal Forces.* Danish Geotechnical Institute. Bulletin No. 12. Copenhagen. 1961.

40. Ng C.F. *Behaviour of Buried Pipelines Subjected to External Loading.* PhD Thesis. University of Sheffield. November 1994.

41. Terzaghi K. *Theoretical Soil Mechanics.* John Wiley & Sons Ltd. 1943.

42. Meyerhof G.G. *Influence of Roughness of Base and Ground-water Conditions on the Ultimate Bearing Capacity of Foundations.* Geotechnique. Vol. 5. 1955.

43. Brinch Hansen J. *A Revised and Extended Formula for Bearing Capacity.* Danish Geotechnical Institute. Bulletin 28. Copenhagen. 1970.

44. Trautmann C.H. & O'Rourke T.D. *Behavior of Pipe in Dry Sand Under Lateral and Uplift Loading.* Geotechnical Engineering Report 83-6. Cornell University. Ithaca. New York. 1983.

45. Das Braja M. *Advanced Soil Mechanics*. Taylor & Francis. 1997.

46. Terzaghi K., Peck R.B. & Mesri G. *Soil Mechanics in Engineering Practice*. John Wiley & Sons. 3rd edition. 1996.

47. Bolton M.D. *The strength and dilatancy of sand.* Geotechnique 36. No. 1. 1986.

48. Carter M. & Bentley S.P. *Corellations of Soil Properties.* Pentech Press Ltd. London. 1991.

49. Lambe T.W. & Whitman R.V. *Soil Mechanics*. John Wiley & Sons. 1969.

50. Rowe P.W. *The relation between the shear strength of sands in triaxial compression, plane strain and direct shear*. Geotechnique 19. No. 1. 1969.

51. AS 2885.1-1997. *Pipelines-Gas and liquid petroleum. Part 1: Design and construction.* Standards Australia. 1997.

C571/025/99

Real-time pipeline leak detection on Shell's North Western Ethylene Pipeline

J ZHANG and **L XU**
REL Instrumentation Limited, Manchester, UK

ABSTRACT

In the past ten years, a number of pipeline leak detection systems have been implemented on various operational pipelines. Unfortunately the feedback from pipeline operators illustrates that some of these systems have not performed satisfactorily for the following reasons:

- They generate frequent nuisance alarms when there is no leak in a pipeline.
- They are difficult for users to understand.
- They are expensive to maintain.

Consequently, there is a tendency for leak alarms to be neglected and in some cases the systems are switched off completely.

This paper examines the application of state-of-the-art technology (ATMOS PIPE) to the 413 kilometre long North Western Ethylene Pipeline, for Shell UK Limited. Following the technical introduction, the paper addresses the performance of the system over the past two years. Experience on this high pressure ethylene pipeline demonstrates that

- It is possible to have a reliable real-time leak detection system.
- An effective leak detection system need not be highly complicated.
- System maintenance cost can be minimised with the use of new technology.

1 INTRODUCTION

The North Western Ethylene Pipeline was commissioned by Shell in 1992. The pipeline transports high pressure (dense phase) ethylene from Grangemouth in Scotland, via Cumbria and Lancashire, to Stanlow refinery in Ellesmere Port. The cross-country pipeline is 413 kilometres long.

The smooth operation of such a long and hazardous pipeline demands continuous monitoring. To achieve high integrity, Shell incorporated considerable details during the design of the pipeline and instrumentation system. For example,

- Cathodic protection for the whole pipeline.
- There are 25 block valves along the pipeline that can be operated remotely.
- Pressure and temperature meters are installed both upstream and downstream of the block valves.
- At the inlet and outlet of the pipeline, two sets of flow, pressure, temperature and density meters are installed.
- Two SCADA computers are used. One works as the duty machine and the other as hot standby.
- Redundant telecommunication systems are used so that it is highly unlikely for the SCADA system to lose communication with two consecutive block valves.
- A pipeline leak detection system.
- Regular intelligent pigging.
- Road and aerial surveillance.

In the past ten years, a number of pipeline leak detection systems have been implemented on various operational pipelines. Unfortunately the feedback from pipeline operators illustrates that some of these systems have not performed satisfactorily (1), (2). The main problem with these systems is the generation of frequent nuisance alarms. To overcome this problem, Shell has developed a statistical leak detection system following several years of research and field trials (3). This system is licensed to REL Instrumentation Limited and it is trade marked as ATMOS PIPE.

ATMOS PIPE had been previously applied successfully to a liquid propylene and crude oil pipeline in Shell UK Limited (4), (5). To achieve the same high level of reliability, Shell decided to install ATMOS PIPE on the North Western Ethylene Pipeline.

ATMOS PIPE was implemented on a Personal Computer in December 1997. It gets all the instrument data from the existing SCADA computer at 30 second intervals. After processing the data, the pipeline status (normal or leak) is sent back to the SCADA computer together with the leak rate and location estimates.

Shell performed a Site Acceptance Test in March 1998. ATMOS PIPE has detected a "leak" of 8 ton/hour (0.38 m^3/minute) in 15 minutes, with very accurate leak size and location estimates.

During normal pipeline operations, ATMOS PIPE does not generate nuisance alarms, and over the last two years of operation, it has proven to be highly reliable.

2 PIPELINE DESCRIPTION

The lay out of the pipeline is shown in **Figure 1**. The pipeline is 10" diameter, and **413 km** long. The pipeline runs through hilly areas with large elevation changes (**Figure 2**).

There are 25 block valves along the pipeline. The locations of these block valves are shown in **Figure 1**. About half way along the pipeline, at Block Valve 12, an intermediate pump station (IPS) is present. The pumps at the IPS were commissioned in 1998, and are started and stopped to suit operational conditions.

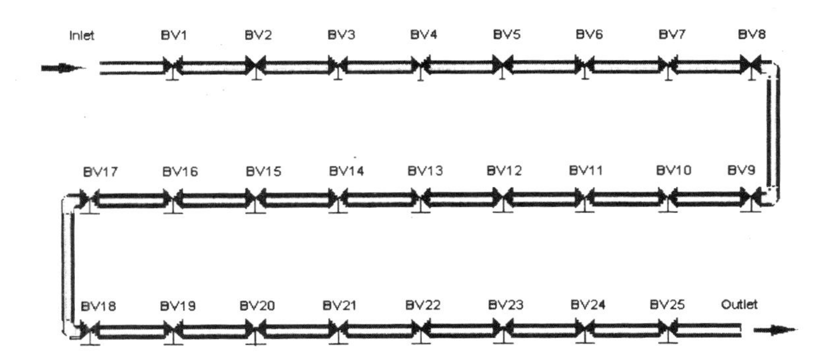

Figure 1 Lay out of the ethylene pipeline

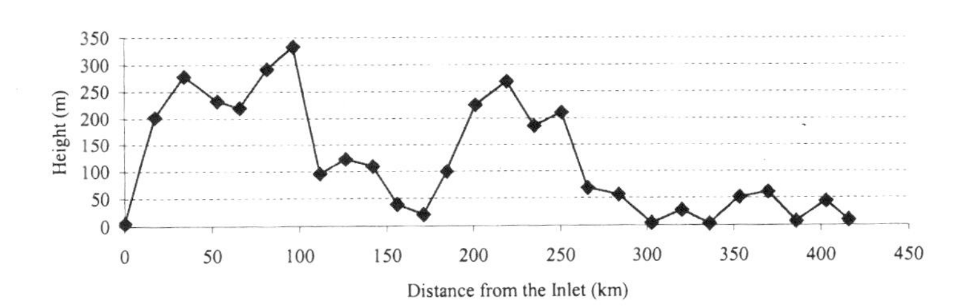

Figure 2 Pipeline elevation profile

Flow, pressure, temperature and density meters are available at the inlet and outlet of the pipeline. Pressure and temperature are measured both upstream and downstream of the block valves. In total there are about 160 measured variables including 25 block valve position indicators. Each of these measurements has a quality status attached, indicating the confidence level in the measurements.

3 IMPLEMENTATION OF ATMOS PIPE

The project was initiated in May 1997. The existing instrumentation system was already collecting all the flow, pressure, temperature, density and valve data. Therefore the only hardware required to install ATMOS PIPE was a Personal Computer. **Figure 3** shows the interface between ATMOS and the instrumentation system.

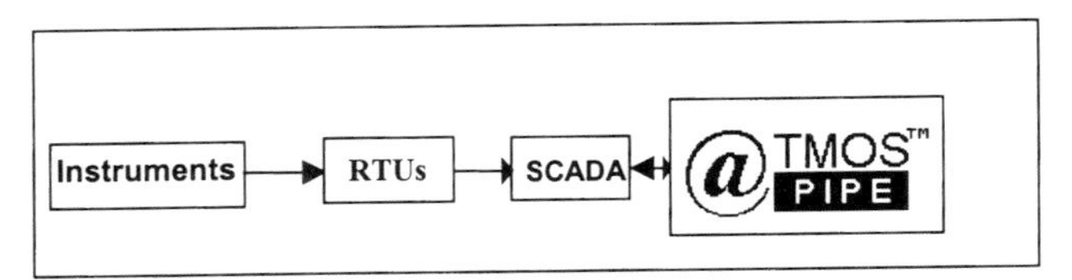

FIGURE 3 THE INTERFACE BETWEEN ATMOS AND THE INSTRUMENTATION SYSTEM
(RTU — Remote Terminal Unit, SCADA — Supervisory Control And Data Acquisition)

In August 1997, the PC was installed on site to collect the operational data. These data were then used to tune the system parameters so that no false alarm was generated under normal operational conditions. Based on the customer's specifications on the leak detection response time and data collected between August and November 1997, ATMOS PIPE was commissioned in December 1997.

The main functionality of the system is:

- Collection of flow, pressure, temperature and valve data at 30-second intervals.
- Validation of the above data so that faulty instruments are diagnosed and "bad" data are rejected.
- Detection of leaks under different operational conditions: transient, steady state and shut-in.
- Estimation of leak size and location.
- Record of historical data and events.

To optimise the system performance, the following key features were included (1):

Reliability
There are 160 measured variables from the SCADA system. It is highly probable that one of these measurements is not correct at any specific time. To eliminate errors introduced by such data faults, a comprehensive data validation module is implemented. This operates in three stages:

- Status check. Each measurement has a status variable attached. If the status indicates that the measurement is not "OK" then it will not be used.
- Redundancy check. Since there is sufficient redundancy in the instrumentation system for ATMOS PIPE to compare different readings, a datum will not be used if it is inconsistent with its redundant measurement. For example, if the pressure readings upstream and downstream of a block valve differ by more than the pressure drop over the valve, then both readings will be suspect. Only the pressure that is consistent with the other block valve pressure measurements will be used.
- Independent check. Each data point is validated and it is rejected if it is out of range, does not change at all for a period of time, or the rate of change is too high within a short time period.

Robustness
The instrumentation system consists of field instruments, telecommunication equipment and SCADA computers. A failure in any of this equipment would result in the loss of measurement

data. ATMOS PIPE was designed so that it will continue monitoring the pipeline as long as the flow measurements at the inlet and outlet are available i.e. even when all the pressure and temperature data from the block valves and inlet and outlet have failed.

To maintain a high reliability level when some of the instruments are not available, the statistical parameters are changed automatically. For example if the measurements from all the even block valves are unavailable, then the system will be desensitised such that the leak detection time will be 10% longer than when all the measurements are available.

Sensitivity
Without loss of reliability, ATMOS PIPE was designed such that the detection time is minimised for various operating conditions and changing instrumentation system scenarios. Three operating modes are included:

steady state,
medium operational change,
large operational change.

Changes in the above operating modes are detected automatically and different sets of statistical parameters are used for each of the modes.

Accuracy
Leak rate and location estimates are provided after a leak alarm is generated. Accurate leak rate estimate is achieved by removing instrument errors using statistical calculations. These calculations are carried out continuously during normal operating conditions. Therefore gradual instrument drifts over a long period of time will be excluded from leak size estimation.

It is difficult to estimate leak location accurately based on instrument readings. The North Western Ethylene Pipeline is particularly hard for a conventional leak detection system because

- the equation of state is not well established for ethylene
- the pipeline elevation varies significantly and it is difficult to measure the elevations accurately (**Figure 2**).

ATMOS PIPE calculates the leak location statistically. Since flow and pressure data are used to calculate the actual pressure profile, no theoretical assumption is made about the equation of state and pipeline elevation data. With the availability of accurate flow and pressure measurements for this pipeline, accurate leak location estimates have been achieved.

4 SITE ACCEPTANCE TEST

Following the installation in December 1997, ATMOS PIPE had worked continuously without generating any false alarms. Shell conducted a Site Acceptance Test in March 1998 to test its performance when a leak occurs.

On the 25 March 1998, a leak of about 8 ton/hour (0.38 m^3/minute) was generated by flaring ethylene at Block Valve 2 (34 kilometres from the inlet). ATMOS PIPE generated a leak warning **11 minutes** after the leak and confirmed the leak 4 minutes after the leak warning. At the same time, an accurate leak location was given which was **33.8 km** from the inlet. The average leak size estimated was 7.6 ton/hour.

Figure 4 shows the field test results including the seven statistical variables "lambda0" to "lambda6" and the threshold value. Leak alarm status and leak size estimate are given in **Figure 5**. **Figure 6** illustrates the leak location estimate.

As shown in **Figure 4**, at time **9:35:00** Lambda0, Lambda1, Lambda2, Lambda3, Lambda4 and Lambda5 started to increase indicating the initial leak size was between 5.0 and 12.5 t/h. At time **9:46:30**, Lambda4 was 5.0 which was greater than the threshold value of 4.6. Therefore a leak warning was generated (Alarm status changed from 0 to 1) i.e. in 11.5 minutes Lambda4 increased from –7.0 to 5.0. A leak alarm was generated at time **9:50:00** (**Figure 5**, alarm status changed from 1 to 2), the leak size estimate was 8.69 t/h and location 33.8 Km from the inlet.

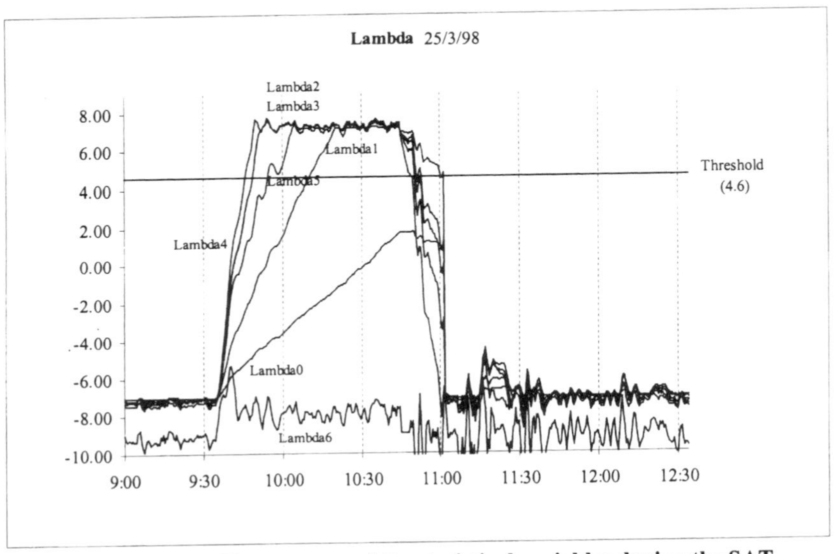

Figure 4 The response of the statistical variables during the SAT

From time 9:50:00 to 10:45:00, the leak size estimate varied between 6.7 and 9.1 t/h and the average is **7.6 t/h** (**Figure 5**). The leak location estimate stayed around 33.8 Km (**Figure 6**).

The leak alarm was cleared (alarm status from 2 to 0) at time 11:00:00 but went back to 1 (leak warning) for one sample before settling back to 0 at time 11:02:00 (**Figure 5**).

Following the successful leak test, Shell has accepted the system and it has been running satisfactorily since the site acceptance test.

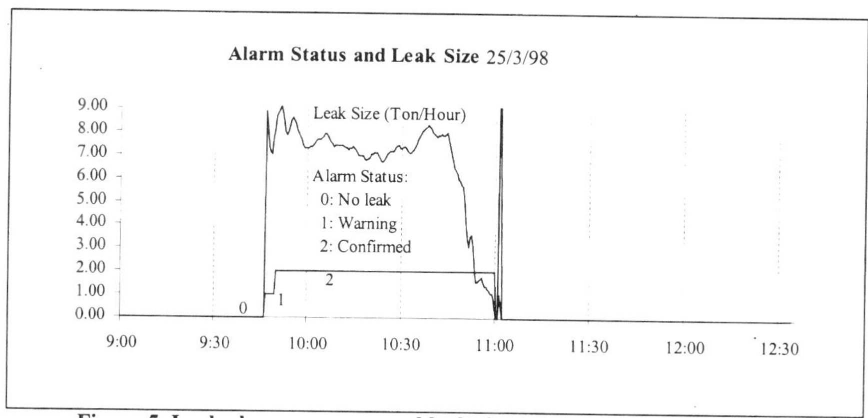

Figure 5 Leak alarm response and leak size estimate during the SAT

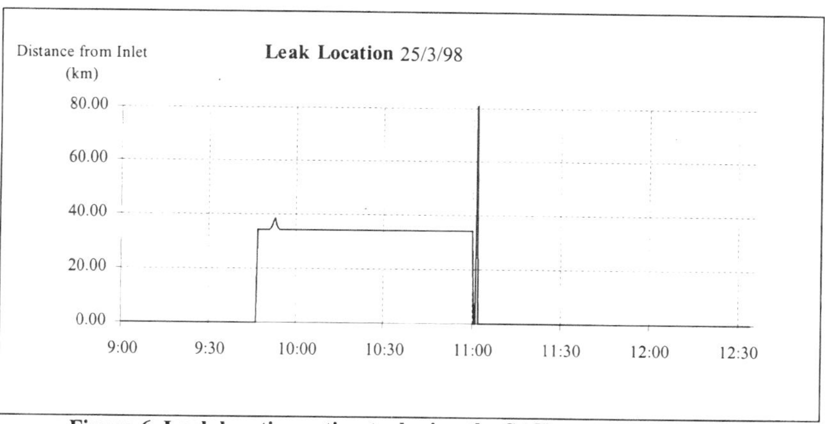

Figure 6 Leak location estimate during the SAT

5 SYSTEM PERFORMANCE DURING NORMAL OPERATIONS

ATMOS PIPE has been monitoring the North Western Ethylene Pipeline since December 1997. Significant operational changes have occurred during this period, for example:

- The IPS pump was commissioned at the middle of the pipeline so that one or two pumps can be started and stopped when required.
- Pigging has been carried out for the whole pipeline length.
- The pipeline was in a shut-in condition for a short period of time when no flow is pumped into or out of the pipeline.
- Partial loss of instrument data due to power failure caused by lightning or other events.

ATMOS PIPE has experienced the above changes without loss of functionality or reliability. In 1998, for example, only one leak alarm was generated due to IPS pump shut down and ATMOS did not recognise the change quickly enough. This type of alarm will not occur again as the IPS status is now sent to ATMOS to inform it of any changes immediately.

In addition to leak detection, ATMOS PIPE has been used to provide instrument monitoring. Both on-line and off-line analysis can be carried out for each individual instrument so that its performance is assessed and additional information is provided for maintenance and support. Other value-added services provided by ATMOS PIPE include:

- Flow discrepancy analysis between the inlet and outlet,
- Inventory tracking and survivability analysis,
- Pig tracking,
- Management information.

Although ATMOS PIPE was installed back in 1997 it is fully year 2000 compliant. There was no additional cost to the customer for Y2K tests. The overall maintenance cost is also low compared with conventional systems.

6 COMPARISON WITH OTHER TECHNOLOGIES

Different leak detection technologies can be used to meet the application requirements (4). For the continuous monitoring of a pipeline, the following software-based methods are available:

- Volume or mass balance
- Rate of change in flow or pressure
- Hydraulic modelling
- Pressure point analysis
- Statistical analysis (ATMOS PIPE).

Depending on the design and implementation of a particular technology, the performance of a leak detection system varies significantly. The best technology may not work if it is engineered poorly. Therefore only a general comparison of performance is given in Table 1 and no particular reference is made to any commercial products.

Table 1 Comparison of different methods

Method	Reliability	Robustness	Sensitivity	Accuracy
Mass balance	Medium	High	Medium	Medium*
Rate of change	Medium	High	Medium	N/A[&]
Hydraulic modelling	Medium	Low	High	Medium
Pressure point analysis	Medium	High	Medium	N/A[&]
ATMOS PIPE	High	High	High	Medium

* Leak location is not estimated by mass balance method.

& Leak size and location are not estimated.

As shown in Table 1, most of the above methods are robust since they can monitor a pipeline continuously even when some instruments fail. However, a common shortcoming of these methods is that they either do not provide a leak location estimate or cannot pinpoint a leak accurately. Although ATMOS PIPE located the leak accurately during the site acceptance test on the North Western Ethylene Pipeline, it is because the leak was at a block valve where pressure measurements were available. For most applications, it is not feasible for any of these methods to pinpoint a leak with no error. The main reason is that these methods work based on measurements given by field instruments that are not 100% accurate. To improve the leak location accuracy, continuous research and development have been carried out at REL Instrumentation Limited.

7 CONCLUSIONS

ATMOS PIPE is state-of-the-art leak detection technology. Its application to the North Western Ethylene Pipeline proves that it has minimum false alarms during normal pipeline operating conditions and it is cost effective to maintain.

The Site Acceptance Test shows that it detected an 8 ton/hour (0.38 m^3/minute) leak in 15 minutes with accurate leak rate and location estimates. The statistical characteristics provide it with a self-tuning capability allowing it to monitor the pipeline for its entire life cycle at minimum costs.

Following the successful application to the North Western Ethylene Pipeline, Shell has installed ATMOS PIPE on both the Runcorn to Stanlow and Stanlow to Montell ethylene pipelines. Recent applications to other gas and liquid pipelines have further proven that it is possible to have a robust and reliable but simple leak detection system.

Acknowledgement
This paper is published with the kind permission of Shell UK Limited, which is greatly appreciated.

REFERENCES

1. API, 1995. "Evaluation Methodology for Software Based Leak Detection Systems", API Publication 1155.
2. M.N. Mears, 1993. "Real World Applications of Pipeline Leak Detection", Pipeline Infrastructure II, Proc. Int. Conf. ASCE.
3. X.J. Zhang, 1993. "Statistical Leak Detection in Gas and Liquid Pipelines". Pipes & Pipelines International, July-August, p26-29.
4. X.J. Zhang, 1997. "Designing a Cost-effective and Reliable Pipeline Leak Detection System", Pipes & Pipelines International, January-February, p20-26.
5. J. Zhang, E. Di Mauro, "Implementing a Reliable Leak Detection System on a Crude Oil Pipeline", Advances in Pipeline Technology 1998, Dubai, UAE

C571/040/99

Risk and integrity management of a transmission pipeline

P HOPKINS
Andrew Palmer and Associates (a division of SAIC Limited), Newcastle-upon-Tyne, UK

ABSTRACT

Regulatory authorities are moving away from prescriptive approaches in pipeline design and operation to *'risk management'* as the safest and most cost effective means of maintaining and improving safety levels in pipelines. Additionally, operators and regulators are recognising the importance and usefulness of *'management systems'*.

This paper introduces the reader to risk management methods, and management systems, and combines these two approaches to produce a complete risk management system, including emergency planning and procedures, which can be used on a pipeline system.

1. INTRODUCTION

Pipelines must provide a safe method of transporting energy, and pipeline operators must ensure that the public, the environment and property are protected from any associated risks.

Most pipeline operators control these risks by complying with their regulatory requirements and national codes, but regulatory regimes are generally 'prescriptive', and will not be adaptable to differing pipelines with differing needs and associated risks. This presents the dual problems of (i) potentially 'missing' new risks, and (ii) creating an inflexible environment for operators to apply new technologies that can both identify and mitigate the key risks.

This paper commences with a general overview of risk management in business today, then explains risk management, and its advantages, in the pipeline world.

1.1 Risk Management in the world today - a general overview

Risk management in industry today is very broad; traditionally, companies have taken a narrow view on this, such as only considering risks to their business that can be insured against, principally in finance and credit management (1). This is due to a historical concern with interest rates, financial failure of customers, exchange rates, etc.. Hence, traditionally, the responsibility has fallen on the finance department.

The modern approach is broader and takes into account wider issues such as customer satisfaction and technology (e.g. technology companies carry the highest risk by having a long product development cycle). Therefore, risk management monitors and analyses all aspects of business risk, and this is why it is being introduced into pipeline operations.

Risk management responsibility always rests with the executives of the company (the 'board'). It starts at an operational level, being part of the day to day running of the company, and should be included in job descriptions.

1.2 The move towards Risk Management in the pipeline world

Regulatory authorities (2-10) are moving away from prescriptive approaches in pipeline design and operation, to 'risk management' as the safest and most cost effective means of maintaining and improving safety levels in pipelines. Risk management recognises that it is not possible to eliminate all risks, and it recognises that the best way to control risks is the analytical and cost effective use of available resources, and not by simply following regulations and codes (10).

This means that the pipeline industry is changing, from prescriptive (some would say 'restrictive') methods of designing and operating pipelines, to 'goal setting'. Therefore, operators should be aware of these new management methods.

1.3 The move towards 'Management Systems'

Pipeline operators and regulators are recognising the importance and usefulness of formal 'management systems'. A Pipeline Management System (11) is expected to be a requirement in all countries of the European Union in the near future (12).

Management systems bring together company organisation and structure, responsibilities, processes, etc., in a single document that is constantly reviewed and audited to quantify its usefulness and effectiveness. A pipeline management system includes the parts of a general management system relating to a pipeline and is the obvious 'home' for risk management methods and procedures.

2. THE INCREASING USE OF RISK MANAGEMENT METHODS

Many countries are actively using, or moving towards, risk management methods.

2.1 UK

The Pipelines Safety Regulations 1996 (5,6) in the UK cover most oil and gas, onshore and offshore, transmission pipelines. These Regulations are not prescriptive, but goal setting, where operators can base design and operation on 'fitness-for-purpose'.

Regulation 23 requires the operator of a major accident hazard pipeline to have available a 'Major Accident Prevention Document' (MAPD). A 'major accident' is defined as 'death or serious injury involving a dangerous fluid'.

The MAPD is a management tool to ensure that the operator has assessed the risk from major accidents and has introduced an appropriate Safety Management System[1] to control these risks. The pipeline MAPD must be supported by a Safety Management System (11, 13, 14) which is in place for the control of the safety of the pipeline throughout its lifecycle. The safety management system should cover the organisation and arrangements for preventing, controlling and mitigating the consequences of major accidents.

2.2 Canada

In Canada, risk assessment and management is being promoted by the Pipeline Risk Assessment Steering Committee, which has developed a Canada-wide database of reportable pipeline incidents and characteristics.

A non-manadatory appendix, 'Guidelines for Risk Analysis of Pipelines', was included in the Canadian Standards Association Standard Z662 in 1996. The latest edition outlines risk assessment concepts and summarises the risk assessment process, and acknowldges both qualitative and quantitative risk analysis (9).

2.3 USA

The USA Office of Pipeline Safety has instigated a risk management programme, where a partnership of industry and regulators is using risk management as a potential method of producing equal or greater levels of pipeline safety in a more cost effective manner than the current regulatory regime (2, 3, 7).

The approach in the USA is to implement a complete risk management programme on a pipeline system, and to make it a requirement for operators to ensure that risk management is integrated with the company's business practices.

2.4 Europe

The European Commission is reviewing the control of 'major accident'[2] pipelines (12) with the aim of controlling major accidents involving pipelines carrying dangerous substances[3].

This will entail requirements on all member states to draw up a Major Accident Prevention Policy (MAPP) that establishes the overall aims and principles of action with respect to the control of major accident pipelines. There will be another requirement for a Pipeline Management System that ensures the MAPP is properly implemented.

---

[1] The UK Royal Society (15) defined 'safety' as 'the freedom from unacceptable risks of personal harm' (see footnote on following page), and 'safety management' as 'the application of organisational and management principles to achieve optimum safety with high confidence'.

[2] A 'major accident' is defined (12) as an occurrence such as a major emission, fire, or explosion leading to serious danger to human health and/or the environment.

[3] A 'dangerous substance' is defined in the Regulatory Benchmark (12). These substances are graded from '1' (very toxic) through '5' ('highly inflammable') to 8 (classes including those that react with water).

3. RISK AND INTEGRITY MANAGEMENT

3.1 What is 'Risk'?

All operators want a pipeline that is safe[4] (does not pose a major risk to the population and environment) and secure (does not pose a major risk to supplies). Therefore, they require high 'integrity'. This is usually interpreted as a low probability of the pipeline failing (or failing to deliver), and is very much a legacy of the 'old' approach of refusing to admit that a pipeline may fail, with significant consequences.

Risk is calculated by combining the likelihood of a hazardous event, with its consequences:

RISK = f(Probability of Event, Consequence of the Event) ... *1*

Consequence analysis will depend on: pipeline type, product, release rate through orifices, toxicity of product, generation and dispersion of vapour clouds and flame jets, thermal radiation hazards, vapour cloud explosions, etc..

3.1.1 How do we currently deal with risk?

Traditionally, transmission pipelines are designed in accordance with design codes such as the American Pipeline Standards ASME B31.4 or B31.8. Most national and international pipeline design codes are based on these ASME codes.

They use 'deterministic' limits, e.g. a design stress of 72% of the specified minimum yield strength (SMYS) is used, based on conservative assumptions such as minimum wall thickness.

Therefore, by limiting stress, we will ensure a low probability of failure - this is one half of the above risk equation 1. The consequences of failure are mitigated by ensuring that we have a low failure probability in the case of liquid lines, but in gas lines we limit consequences by limiting the number of people (buildings) in the vicinity of the pipeline.

3.1.2 How can we link 'Risk' and 'Integrity' Management?

Traditionally, operators have considered pipeline integrity as dealing with failure probability, and pipeline safety as dealing with failure consequences. Therefore, operators will often have 'pipeline integrity management systems' and 'safety management systems' dealing with these two separate aspects. However, we will combine these two aspects into one approach in this paper - 'risk management'.

3.2 What is Risk (and Integrity) Management?

Risk management has been defined by the USA Office of Pipeline Safety as (3):

'a comprehensive management decision support process, implemented as a program, integrated through defined roles and responsibilities into the day-to-day operations, maintenance, engineering management, and regulatory decisions of the operator'.

We will now consider 'risk management' as including both risk (assessment and control) and integrity management aspects (see also Section 5.2 and Table 1, later) and covering

[4] USA Office of Pipeline Safety (3) considers 'safety' as 'the protection of the public, the environment and property' and 'risk' is 'any threat to achieving these goals'.

three key areas (3): Risk Assessment (Analysis), Risk Control and Decision Support, and Performance Monitoring and Feedback

3.3 How can I develop a full Risk Management Programme?

This paper is going to outline a full risk management programme in the form of a system. All the elements of the above risk management programmes and regulations will be contained in the Pipeline Management System (see later), whose task is to ensure that the whole programme is implemented, measured, reviewed, audited and improved.

It should be emphasised that we are only dealing with the risk management system in this paper. The risk assessment element is key and central to its success and the safety of the pipeline. Guidance on risk assessment can be found in the literature (e.g. 16, 17).

The risk management model we will use is detailed in Figure 1. It brings together all safety aspects of our pipeline system.

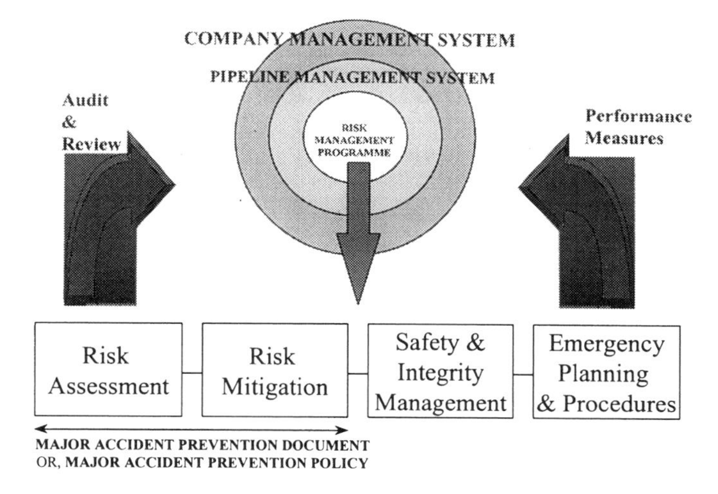

Figure 1. An overview of a complete Risk Management 'Programme', contained within a Pipeline Management System.

We will first describe management systems that can contain and implement our risk management programme and methods. Then we will outline risk management techniques.

4. WHAT IS A MANAGEMENT SYSTEM?

Pipeline operators are now producing formal management systems. Their importance cannot be over-emphasised, and can be illustrated by reference to a recent report dealing with serious allegations of failures in management, management systems, audit, document and quality control, that could all have been avoided with the use of a robust management system (18).

A Management System is a management plan, in the form of a document, that explains to company staff, customers, regulatory authorities, etc., how the company and its assets are: managed, the company targets and goals, how policies are implemented, how performance is measured, and how the whole system is reviewed and audited.

Examples of why a company needs management systems, and the need to constantly review all management procedures, can be found in the literature (e.g. 18, 19).

4.1 Simple Management Systems

An example of a simple management system is shown in Figure 2 (14). In this case we are considering a management policy (safety), and we want to ensure that this policy is robust, well managed (with clear responsibilities and lines of responsibilities) implemented effectively, and in line with stated performance measures.

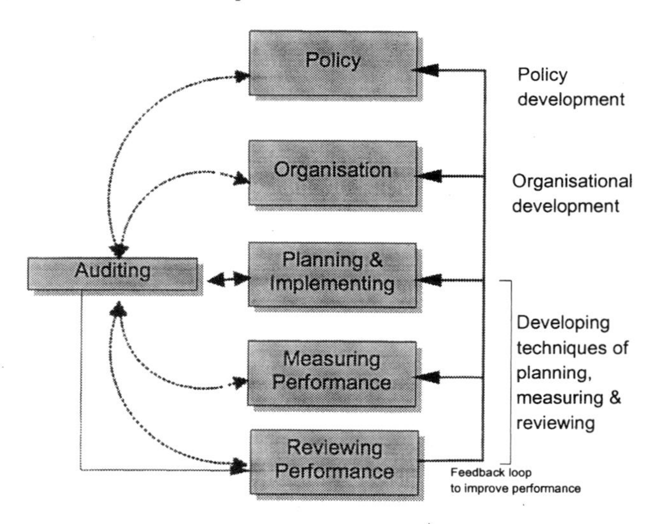

Figure 2. Key elements of a Management System (14)

4.2 Pipeline Management System

A Pipeline Management System is shown in Figure 3 (11, 13, 20, 21). This is a suggested format; different companies require different formats, and different priorities. Each arm of the system in Figure 3 should contain all the elements of Figure 2.

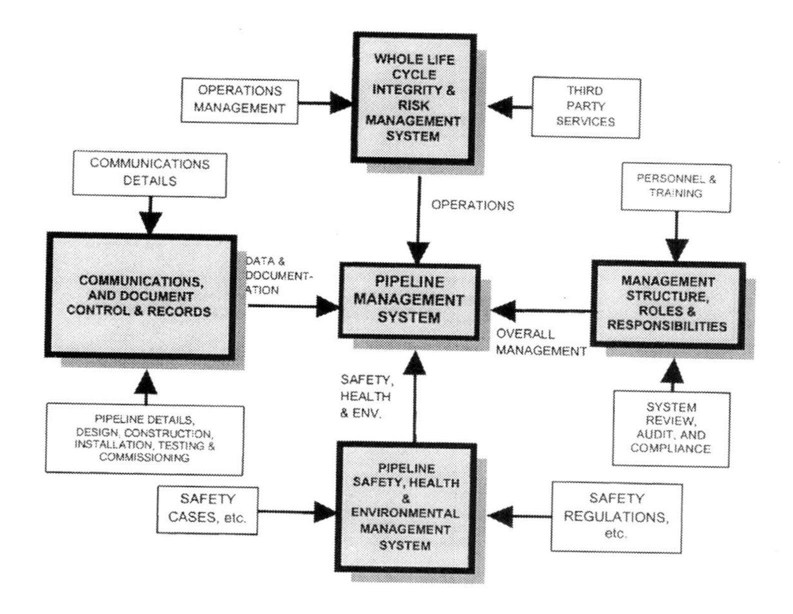

Figure 3. A Pipeline Management System (11)

5. PIPELINE RISK MANAGEMENT SYSTEMS

A risk management system will take the form of a document that clearly sets out missions, responsibilities, policies and procedures, and performance measures. The main elements of this management system are summarised below.

5.1 Overall Risk Management System

The European, UK and USA risk and safety requirements (2, 3, 5, 12) can be used to construct a Pipeline Risk Management System, Table 1. The risk management system brings together all aspects of our pipeline safety and integrity methods.

5.2 Performance measures

Performance measures can be difficult to develop and agree with Regulatory Authorities. But in general, three measures can be proposed (3):

i. *Safety & Reliability* – For example, how are incident data changing
ii. *Resource Effectiveness* – For example, are required resourcings being reduced due (directly) to the management programme?
iii. *Communication & Partnership* – For example, do all stakeholders (operator, regulator, customers, public) appreciate all risks, and agree?

Table 1. A Risk Management System

1. Introduction: Purpose, Objectives, Goals, Company Mission Statements, Corporate Administration.
2. Description of Pipeline Systems & Legal and Statutory Duties: Interfaces with Other Operators' Facilities or Pipelines, Description of design, construction, etc., Standards, and Formal Statement of Legal and Code Compliance.
3. Organisation & Control: Structure, Accountabilities and Responsibilities, Organisation, Relationships with Other Groups/Bodies, Integration into Company and Pipeline Management Systems.
4. Key Personnel - Roles, Responsibilities, Qualifications, Training and Updating of Skills.
5. Stakeholders: Listing of all Stakeholders, with interests and concerns. Information to be given to all stakeholders (pipeline operator details, system details, nature of the major accident hazards, procedures (including communications) in case of an accident, and information concerning emergency plans and procedures).
6. Documentation and Communication Systems: Type, Methods, Location, Feedback from/to all personnel and Stakeholders.
7. Management of Change: both change of management and engineering detail.
8. Risk Management - Analysis, Evaluation and Control for the Whole Life of the Pipeline: Methodology (with limitations and assumptions), types of hazard/risk, adverse events, likelihood, frequency, consequences, quantification, sensitivity and uncertainty analysis, acceptance levels. Identification of major risks, control of risks, evaluation of control options, prevention and mitigation methods, acceptance levels. Emphasis on identifying all major accident scenarios, and their probability, the events triggering the accident, and the extent, severity and consequences. All hazards should be eliminated or reduced to as low *as is reasonably practicable*[5] Where hazards cannot be eliminated, appropriate measures (protection) should be applied.
9. Integrity Management - control, maintenance, inspection & monitoring- policies, procedures & specifications: i. Internal Erosion, Corrosion and External Damage - Control, Mitigation, and Monitoring, ii. Pipeline Geometry, Leaks, Ground Movement - Control, Mitigation and Monitoring, iii. Pressure and Overpressure Control - Control and Monitoring, iv. Definition of Reportable Incident/Damage, Incident Investigation and Analysis, v. Full Listings of all Integrity Monitoring Procedures, Intervals, Responsibilities, Agency Carrying Out Duties, Reporting, vi. Repair, Modification Procedures and Methods, Summary of Emergency Plans & Procedures (see Items 10 and 11), including Liaison with Other Services, viii. Information & Documentation Relating to Pipeline Integrity.
10. Emergency Planning: Responsible persons, co-ordination/liaison/interface roles, actions to control/limit consequences of an accident, early warning systems, information/communications, training by/with local emergency services.
11. Emergency Procedures: Detail procedures, roles and responsibilities, testing, updating, links and compatibility with local services and procedures.
12. Performance Measures: i. Stated Performance Measures, ii. Evaluation of Measures, iii. Improvement Process
13. Management System Review - Responsibility and Frequency
14. Management Review - Responsibility and Frequency
15. Audit of All Elements and Processes - Feedback and Change Implementation.

[5] **'As low as reasonably practicable' (ALARP).** ALARP is achieved by basing criteria on a robust, consistent methodology contained within an appropriate safety assessment philosophy or management system. Hazards which can be reasonably minimised or eliminated cannot be considered acceptable, however minimal the frequency or consequences appear to be (14).

5.3 'Acceptable' risks and failure probability

A difficult aspect of the risk analysis is setting 'acceptable' levels of risk. There is some guidance in the literature; for example the UK works to 'as low as reasonable practicable' (see footnote in Table 1) when considering risks to people, Figure 4.

'Acceptable'[6] failure probabilities are also published in Canadian and Norwegian pipeline standards (see Reference 21 for a summary).

Cost benefit analysis can be used to balance the benefits of reducing risks against the costs incurred for a particular option for managing the risks. These analyses will usually be undertaken in the 'tolerable region' in Figure 4. A 'benchmark' benefit for the prevention of death is quoted in the UK as £902,500 (1998 prices) in Reference 10.

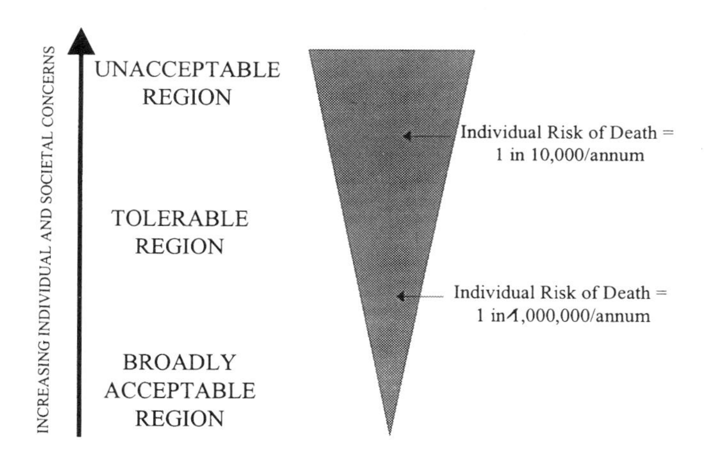

Figure 4. A Conceptual Model for the Tolerability of Risk (of physical harm, etc.) to the UK Population (10)

5.4 Emergency Planning & Procedures

Figure 1 shows the key role of Emergency Planning and Procedures in our risk management system. References 5 and 6 give us a good summary of the requirements for emergency procedures and planning. Reference 13 summarises their content, and format.

5.5 Review and Audit of the System (11, 13)

The whole risk management system should be regularly **reviewed** against specified criteria. Finally, the whole system, and all its elements should be subject to regular **audit**. This audit should not be a '*compliance audit*', that merely checks that the paperwork is in place, and

[6] It should be noted that 'acceptable' and 'tolerable' risks are not necessarily the same. Reference 10 considers 'tolerable' "*as a willingness to live with risk so as to secure certain benefits and in the confidence that the risk is one that is worth taking and that it is being properly controlled*". It considers 'broadly acceptable' as "*the 'level of residual risk regarded as insignificant and further effort to reduce risk not likely to be required as resources to reduce risks likely to be grossly disproportionate to the risk reduction achieved*".

the staff are following all the policies, etc.. The audit should be a *'critical audit'*, that appraises both the system in place, and each element, to ensure it is performing its stated function effectively.

5.6 The way ahead

The previous sections have outlined a complete pipeline risk management system. It is now a relatively easy step to implement this system in a pipeline company.

The preferable way to do this is to follow the above steps (Table 1), and produce a 'paper-based' system. At the core of this system will be a risk analysis (Item 8, Table 1). This analysis can be either quantitative and/or qualitative. Later versions of the system could be in the form of software.

6. CONCLUSIONS

This paper has covered two major new applications for pipeline operators: risk managment methods, and management systems.

It has shown how the risk management methods and recommendations from North America and Europe can be combined to create a comprehensive 'risk management system'. This system includes all aspects of the integrity and safety management of a pipeline, and is positioned within an overall pipeline management system.

A risk management system can be constructed using the guidelines detailed in this paper. Pipeline operators are strongly recommended to produce these risk management systems, as they provide benefits, and are being adopted by the industry and regulatory authorities.

REFERENCES

1. Sunday Telegraph, UK, Jan 31 1999.
2. Anon., 'Pipeline Risk Management', Newsletter of the Office of Pipeline Safety, DoT, USA, Vol. 3, Issue 1, May 1998.
3. Anon, 'Risk Management Program Standard', Office of Pipeline Safety, DoT, USA, September 1996.
4. Reid, R. J., 'World's Pipeline Industry, While Safe and Reliable, Must Learn to Cope With Change', Oil and Gas Journal, August 1998, p.28.
5. Anon., The Pipeline Safety Regulations 1996 (SI 1996 No. 825), HMSO, UK, 1996.
6. Anon., 'A Guide to the Pipeline Safety Regulations, Guidance on Regulations', L82, HSE Books, HMSO, UK, 1996.
7. Leewis, K. J., Shires, T., 'Progress of the US Department of Transportation Risk Management as a Regulatory Alternative', International Pipeline Conference - Volume 1 ASME 1998, p.41.
8. Cicansky, K., Yuen, G., 'Risk Management at TransCanada Pipelines', International Pipeline Conference - Volume 1 ASME 1998, p.9.

9. Anon., 'Oil and Gas Pipeline Systems', Z662-99, Canadian Standards Association, 1999, and Stephens[7], M., CFER, Canada, Private Communication to P Hopkins, June 23 1999.
10. Anon., 'Reducing Risks, Protecting People', Health and Safety Executive Discussion Document, DDE11, HSE Books, 1999.
11. Hopkins, P., Cosham, A., 'How Do You Manage Your Pipeline?', 7th Int. Conference on Pipeline Risk Management and Reliability, Houston, USA, November 1997.
12. Anon., 'Regulatory Benchmark[8] for the Control of Major Hazards Involving Pipelines', (Draft) Pipelines Safety Instrument, European Commission, Italy, July 1998.
13. Hopkins, P., 'Risk And Integrity Management Of A Transmission Pipeline', 2nd Int. Conf. on 'Advances in Pipeline Technology '98', Dubai, UAE, IBC, October 1998.
14. Anon., 'Successful Health and Safety Management', Health and Safety Executive Books, HS(G) 65, HMSO, UK, 1995.
15. Anon., 'Risk Analysis, Perception, Management'. The Royal Society, London, UK, 1993.
16. Hopkins, P., Hopkins, H., Corder, I., 'The Design and Location of Gas Transmission Pipelines Using Risk Analysis Techniques', Risk and Reliability Conference Aberdeen, UK, May 1996.
17. Acton, R. A. et al, 'The Development of the PIPESAFE Risk Assessment Package for Gas Transmission Pipelines', International Pipeline Conference, Vol 1, ASME 1998, p.1.
18. Gillard, M., Rowell, A., Jones, M., 'Oil Pipeline Disaster 'Imminent'', The Guardian Newspaper, 12th July 1999, p1.
19. Anon., 'Investigation into the King's Cross Underground Fire', HMSO, London UK, 1988. Also 'Management by Memo Led to Fundamental Errors', The Guardian Newspaper, UK, 11th November 1988.
20. Hopkins, P., 'New Design Methods for Quantifying and Reducing the Number of Leaks in Onshore and Offshore Transmission Pipelines, Conf. On 'Leak Prevention of Onshore and Offshore Pipelines', IChemE., London, May 1997.
21. Hopkins, P, Lamb, M., 'Incorporating Intelligent Pigging Into Your Pipeline Integrity Management System', Onshore Pipelines Conference, Berlin, Germany, 8-9th December 1997.

[7] Member of CSA Risk Task Force

[8] A 'Benchmark' serves as a basis for self-assessment performed by Member States to compare their existing legislation.

C571/026/99

The application of risk-based approaches to common pipeline issues

B MORGAN and **D HEATH**
Genesis Oil and Gas Consultants Limited, London, UK

ABSTRACT
This paper describes the main elements of a risk-based methodology that can be applied across a broad range of situations to optimise the cost-effective operation of ageing pipelines. Examples are given where this approach may prove beneficial. The importance of the integration of such an approach into an overall pipeline management system is discussed.

1 INTRODUCTION

The current business climate has led to an increasing need for pipeline operators to:

- Extend the life of pipelines
- Uprate them in terms of pressure
- Change their use (different fluids) or
- Tie-in new pipelines to much older systems
- Optimise inspection frequencies

The technique of risk assessment whether qualitative or quantitative (QRA) has long had an application in the safety assessment of pipelines primarily at the design stage, for land use planning or to determine the need or otherwise for sub-sea isolation valves (1). However, the same basic methodology can be applied usefully in many other situations in order to support explicit decision-making, and with a much broader interpretation of risk. In the context of this conference, these include:

- Risk based inspection:
  - Enabling targeting of those pipelines it is necessary to inspect
  - Identifying those pipeline features which are the main concerns
  - Assessing the significance of different risk factors
  - Enabling demonstration that extending operating life is viable

- Risk based justification of uprating or other change of use:
  - Will the likelihood of failure change?
  - Are the consequences different?
  - Are the risk differences acceptable from a safety, environmental and commercial viewpoint?

- Risk based arguments for justification of tie-in of new pipelines to ageing pipelines by providing a common reference measure to evaluate the impact of:
  - Different codes
  - Different pressure ratings
  - Different material specifications

This paper seeks to describe how a simple risk based methodological framework can be used to address all these issues as part of an overall pipeline management system.

2 METHODOLOGY

Regulations internationally are becoming much more risk based, relying on a goal-setting framework rather than a prescriptive requirement. This allows operators a far greater degree of freedom to manage their own business risks in the most appropriate and cost-effective way. However, there is still a reluctance to adopt something other than an "off-the-shelf" specification, due to the difficulty in operating within environments where the boundaries are less well defined.

In discussing the question of risk, it is important to remember that in deriving codes and standards of the more traditional kind, risk assessment was an implicit part of the process. This is because in deriving design factors, some internal thought processes produced a result, which individuals considered represented a reasonable level of protection against failure without being unduly cautious. An explicit risk assessment therefore exposes thought processes and provides an audit trail for justification. It also makes the analysis of sensitivity to different factors easier to investigate and interpret.

Assessment of risk involves a consideration of how likely something is to occur and its consequence and is often written as:

$$\text{Risk} = \text{likelihood x consequences}$$

The consideration of both aspects is important as it avoids concentration on e.g. very catastrophic possibilities, which are extremely unlikely.

If the process of risk assessment is to be useful it is important that both the list of what can go wrong and the whole range of consequences are as complete as possible. In the case of pipelines however this is simplified because the undesirable consequence will normally be a failure of some kind leading to a loss of containment. There may be a number of ultimate consequences of concern from a loss of containment however, including the following categories:

- Safety
- Environmental
- Loss of production revenue
- Contractual liabilities
- Repair costs
- Loss of public image

The relative importance of each of these will vary under different circumstances e.g. the location of the failure will greatly affect the safety/environmental implications, as will the nature of the fluid (oil vs. gas — sweet, sour etc.). All of these consequences, with the possible exception of public image can be measured to some extent in terms of financial loss. Safety measures are commonly subjected to a cost benefit approach to demonstrate the ALARP (As Low As Reasonably Practicable) principle where the risk reduction is weighed against the implied cost of averting a fatality through its implementation. Environmental impact can be difficult to attribute costs to but an approximation may be made by using the clean up costs, both those immediately incurred and any longer term effects depending on the sensitivity of the local habitat.

The three steps below describe a possible approach to the risk assessment process. Examples of how these may be applied in different situations are given in the next section.

STEP 1 - WORKSHOP / BRAINSTORM SESSION

This is divided into two principal parts:

- Identification of potential failure causes
- Assessment of the likelihood and consequence of the failures

This session will be led / facilitated by a chairman and recorded. It is important to strike the balance between having all relevant areas of knowledge represented, whilst keeping the number to a manageable size and focussed.

In this session a typical way of semi-quantifying the risks is by the use of a risk matrix such as that shown below. It may be necessary to develop some basic acceptance criteria to enable some initial screening of importance.

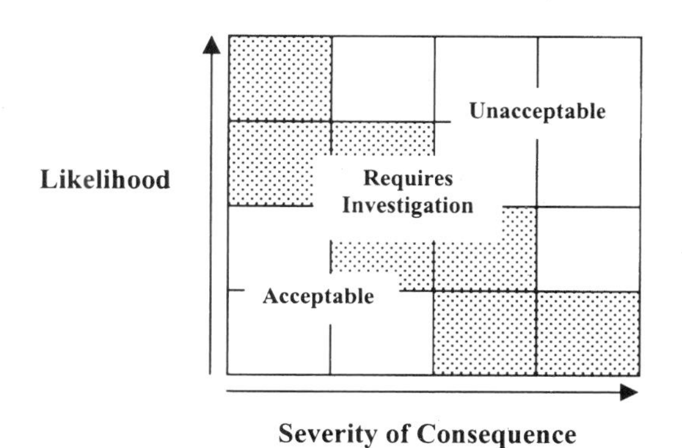

Figure 1 – A Typical Risk Matrix

As there is a range of consequences with different likelihoods, it may be necessary to plot these separately on the chart. The axes may be defined in a number of ways. The consequence axis may be purely in terms of rough cost, or for safety in terms of injuries and fatalities, or, for environmental consequences, size of spill. Use of equivalent cost simplifies this aspect by reducing the consequence to one parameter.

For the likelihood axis, it is often helpful to relate the time period to one the participants can relate to, such as:

Likely	-	expect to see this once or twice / year
Unusual	-	expect to see this once in the project lifetime
Improbable	-	may see this once in a human lifetime
Rare	-	unlikely to see this once in a human lifetime

This approach may be used to evaluate differences between two operating/inspection strategies or to investigate how changing a variable may affect the risk. The aim is always to investigate how the situation may be shifted from the highly undesirable upper right corner, to the acceptable lower left corner of the matrix.

This process may be used to screen out low risk items, identify 'must-do's' and then prioritise the "grey" areas. In the context of risk-based inspection, this may be sufficient to determine which pipelines need to be inspected. In the case of e.g. increasing the operating pressure in a pipeline, this would assist in determining whether there was a significant risk involved as shown by a shift on the position on the matrix.

As stated above, the process could end at this point. However, ideally the results would be passed to stage 2 of the process, which would involve some form of quantification.

STEP 2- RISK QUANTIFICATION

This can be either a relative process or an absolute one; there are clearly degrees of quantification possible both for likelihood and consequence. In practice in the context of pipelines more attention is usually given to quantifying the likelihood of failure occurring (i.e. preventing it). This is particularly true in the case of onshore pipelines where it is not practicable to provide mitigatory measures such as isolation valves to a degree (closeness of spacing) which can greatly affect the magnitude of the accident. Conversely for offshore pipelines it is practicable to provide Riser ESD valves, and occasionally, sub-sea isolation valves that limit the inventory sufficiently to affect the outcome. Nevertheless, the principle of "prevention is better than cure" still prevails.

The derivation of relative risk may involve no more than a refining of step 1, possibly by using a better-defined matrix and a numerical scoring of the likelihood and consequence. In particular this can be used to examine the possible change in risk and it may be sufficient to demonstrate that an item has been shifted from the "grey" area to the acceptable through the adoption of a particular measure.

Full quantification involves postulating a failure rate for the pipeline. This may begin with the use of published historical statistics on failure such as those published by Concawe (2), EGIG (3) and the US DoT (4) for onshore pipelines, and PARLOC *(5)* for offshore pipelines. It is important that the failure data are broken down into both the failure mode and the size of release (and if possible, the pairing of these two factors - though this information is often difficult to obtain). This is because the adoption of different strategies will focus on particular failure mechanisms, for example the implementation of intelligent pigging might be expected to reduce the failure rate due to corrosion but not the failure rate due to 3 Party damage. Conversely some risk reduction strategies may decrease one failure mode but increase another e.g. sleeving of pipelines may protect against damage at crossings but may also increase the potential for corrosion failure. Tables 1-4 below give some example summaries of data from these sources.

Table 1 - Summary of Failure Frequencies by Data Type and Wall Thickness (Failure Frequencies/i 000km years) from EGIG and CONCAWE (as derived in Ref 13)

Failure Frequency Data Source	**West Europe CONCAWE (overall)**			**CONCAWE UK only**			**EGIG**		
Wall Thickness range (mm)	< 5	5 to 10	> 10	< 5	5 to 10	> 10	< 5	5 to 10	> 10
Mechanical failure	0.143	0.143	0.143	0.102	0.102	0.102	0.194	0.194	0.194
Operational	0.047	0.047	0.047	0.036	0.036	0.036	0.068	0.068	0.068
Corrosion	0.085	0.085	0.085	0.060	0.060	0.060	0.114	0.114	0.114
Natural	0.013	0.013	0.013	0.009	0.009	0.009	0.017	0.017	0.017
External impact	0.445	0.132	0.015	0.316	0.093	0.011	0.602	0.177	0.020
Total	**0.733**	**0.420**	**0.303**	**0.523**	**0.300**	**0.218**	**0.995**	**0.570**	**0.413**

Table 2 – Summary of EGIG Failure Frequency Data by Failure Size

Failure Type	**Failures/1,000 km.years**				
	Leak	**Hole**	**Rupture**	**Total**	**Percent**
External Interference	0.070	0.170	0.050	0.290	50
Construction Defect	0.070	0.030	0.010	0.110	19
Corrosion	0.080	0.002	0	0.82	14
Ground Movement	0.010	0.012	0012	0.034	6
Hot Tap in Error	0.020	0.006	0	0.026	5
Other	0.030	0.003	0	0.033	6
Sub Total	0.280	0.223	0.072	-	-
Total	-	-	-	**0.575**	**100**
%	48.70	38.78	12.52	100	-

Table 3 – US DoT Failure rate Frequencies (per 1,000km years)

Pipeline system Failure Mode	US Gas Line Data 1970 – 80 (Raw Data)	US Gas Line Data 1970 – 80 (Amended)*
Defect	0.13	0.046
Corrosion	0.20	0.033
External Impact	0.31	0.165
Environment	0.11	0.015
Operational/Other	0.04	0.014
Total	**0.79**	**0.273**

* For the case of a 10" line and when old pipelines and those without corrosion protection are removed.

Table 4 – Loss of Containment Data for PARLOC 96 (5)

Diameter of Pipeline (inches)	Riser Failure Frequency (per riser year)	Platform Safety Zone Failure Frequency per 500m per year	Mid Line Failure Frequency per 1000 km years
2 to 9	1.14 e-03	5.14 e-04	0.201
> 10	1.04 e-3	6.42 e-04	0.012

These sources of data give useful generic failure data in the absence of other information. In each case however the generic figures may be substituted for more accurate data from a company's own records based on actual failure information, or from e.g. corrosion assessments as a result of intelligent pigging or other surveys. Alternatively, a probabilistic fracture mechanics approach may be used to derive a failure probability. The level of detail that is employed depends on the importance of getting a more precise answer, and therefore the amount of effort it is considered worth expending to obtain this greater accuracy. These kinds of assessment may be commonplace and relatively readily available in many pipeline-operating companies. This information may be used to both form a baseline current situation assessment and then to predict how a failure mode will be affected by a change in practice such as increased pigging frequency, increased pressure etc. The equivalence in risk terms of different possible risk reduction measures or operational strategies can be also be assessed in this way.

In the absence of in-house information or more sophisticated approaches, another approach may be used by which the generic failure rates are modified by means of expert judgement, upwards or downwards. So for example, it could be argued that regular intelligent pigging may reduce the likelihood of a rupture of the line due to corrosion or some other detectable slow growing defect to close to zero, whereas for lesser leaks these may be reduced by, say *75%*. This is common practice in safety risk assessments where quantitative changes are argued on a qualitative engineering judgement basis.

It is important to remember that the approach is the same whether there is a high degree of detail or whether there is dependence on generic failure rates; set the current baseline failure rate and then investigate the effect on that failure rate by the measure/change in operation. It should be noted however, that it may be more difficult to gain acceptance of an approach by the authorities that makes no reference to the generic, published failure data, particularly if spectacularly low failure rates are predicted. This is because some theoretical approaches tend to ignore the human element and their part in every failure, whereas historical data implicitly include this. It should be remembered in this context that the human element could just as easily the operator of a calculator as a forklift truck driver!

STEP 3- COST BENEFIT ANALYSIS

Once the potential benefit of particular strategies have been identified, whether they are implemented or not will depend on a number of other factors, however, this will undoubtedly include an element of their relative cost-benefit to the business. Figure 2 below illustrates the basic methodology involved. Again the methodology remains the same regardless of the degree of detail and more accurate quantification involved. We again see the use of a kind of matrix/plot which immediately highlights good and bad solutions by virtue of their relative position on the chart.

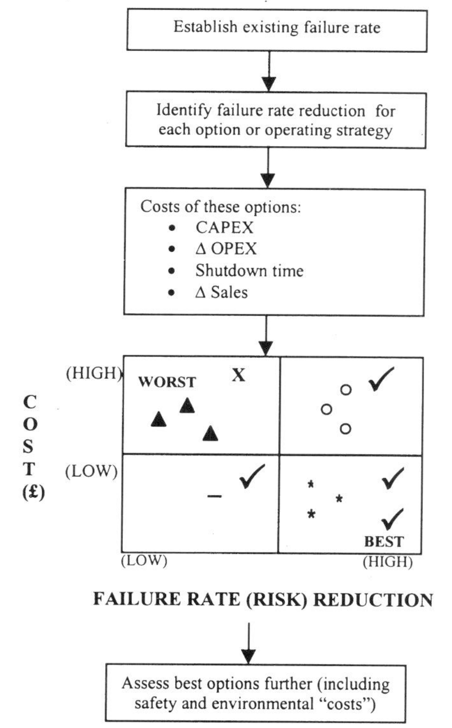

Figure 2 – Cost Benefit Methodology
(Acknowledgement: Based on a figure by David McKnight, Genesis)

3 EXAMPLE APPLICATIONS

There are many examples where a risk based argument has been used, but there are many more where such an approach may well have been cost-effective to the operator. Some examples of potential or actual application of risk arguments are discussed below.

3.1 Demonstration of the acceptability of continued operation of an ageing pipeline

A risk assessment was carried out on a very old 400km long gasoline pipeline in Western Europe. The company was considering deviating the pipeline around some towns. The. company wanted to assess the safety of these people in close proximity to the pipeline to determine the extent to which this deviation was necessary. In addition, the company was concerned regarding its environmental liabilities if a leak were to occur. This was particularly the case where it passed through some relatively remote and inaccessible regions where a leak might go undetected for some time, and also the cost of mobilising a repair would be high.

A safety and environmental risk assessment was therefore conducted; the study is discussed more fully elsewhere (13). The safety study identified only a very small number of areas where some risk reduction was advisable - these were where either a large number of people were simultaneously at risk (a sports stadium) or where there were vulnerable people at risk, such as the elderly, the infirm or children. In these areas, relatively low cost measures such as concrete slabs were sufficient to reduce the risk to negligible levels rather than the more radical and expensive re-routing. The environmental risks were assessed using the same failure frequencies as input but the consequences were assessed by means of the sensitivity of the local habitat and ranked. Some very different habitats were thus assessed to be of a similar sensitivity such as water extraction points for drinking water sources, regions of viticulture, public amenity areas and so on. These rankings were then linked to a clean up cost. Both these risks were expressed visually on a GIS (Geographical Information System) atlas of the whole pipeline route. Having these superimposed on a map enabled the pipeline operators to quickly establish the areas of highest priority to render the existing pipeline suitable without constructing extensive new sections of pipe.

3.2 Uprating existing high pressure gas pipelines

The need to increase throughput both onshore and offshore has led to the requirements of previous pipeline codes being challenged, particularly with respect to design factor/wall thickness requirements. Greater pressures will of course potentially increase consequences of failure due to longer flame lengths, greater cloud travel distances etc. however the impact of these increases may in practice be negligible in terms of the likelihood of fatality or numbers of fatalities. In the event that this is the case, the emphasis turns to demonstrating that raising the pressure does not increase the likelihood of failure (particularly a catastrophic full-bore failure in the case of onshore pipelines). In some cases this may mean demonstrating that strict adherence to existing codes (e.g. the onerous 0.3 design factor in suburban areas required by BS8010 ref. 14) does not in fact have any significant effect on the likelihood of failure, so that the overall risk impact is very small.

The argument on the relaxation of 0.3 rests on an understanding of the leak-before-break behaviour of pipelines, which has been covered extensively in the literature (6, 7, 8, 9, 10, 11, 12). This illustrates an important aspect of this approach, which is that it can force one to re-think through the problem from first principles rather than reach for a code "off the shelf" and accept its prescription.

3.3 Tie-in of a new pipeline to an existing system

Similar arguments to those in 3.2 can be made for tie-in of new pipelines to existing pipeline systems, requiring higher pressures of the ageing pipelines than for which they were designed. A further factor that can be important in this instance is the use of actual manufactured wall thickness and in-service history on incidents and from pigging.

4 PIPELINE MANAGEMENT SYSTEMS

The risk-based approach described here should form a fundamental part of an operator's pipeline management system to assure the integrity of the pipeline, and as such be fully integrated into business processes. The UK Pipeline Safety Regulations (PSR) require a safety management system (SMS) as an essential part of the Major Accident Prevention Document (MAPD) for major accident hazard pipelines. However, the SMS should be a key element of a much broader management system addressing many of the issues we have discussed here.

The UK is not alone in stressing the importance of management systems. The EC has recently been reviewing the necessity or otherwise for an EC Directive on major accident hazard pipelines, which are not covered by existing safety/environmental EC legislation. They have held workshops to discuss the possible approaches with the industry. One of the favoured approaches to emerge was essentially self-regulatory, by means of having suitable management systems in place and the setting of appropriate performance standards.

This approach has also gained favour in the US where a risk management approach to pipeline regulation is being proposed by the Office of Pipeline Safety (OPS). To counter the concern operators may have about deviations from code requirements in the litigious US, the OPS has set up a number of demonstration projects with operators, to run over a number of years. In these projects the regulators, the OPS, and others form an integral part of the project team to work with the operators to define the projects and in particular to develop appropriate performance measures by which the success of the approach may be monitored. Many of these projects (their descriptions are available for review on the OPS web page at http://www.cycla.com/opsiswc) deal with issues arising from the operation of ageing pipelines, and provide interesting reading.

5 CONCLUSIONS

This paper has sought to demonstrate how risk assessment can be a practical tool on the operational phase of pipeline systems. Its applicability is wider than its traditional use in safety assessments and design. It should form an integral part of a pipeline management system to ensure a cost effective approach to asset integrity.

6 REFERENCES

1. Morgan B. (1989), Risk Analysis — A Method for Assessing the Safety of Pipelines, Pipelines Industries Guild Journal 109, 12-19
2. CONCAWE "Performance of oil industry cross-country pipelines in Western Europe" (Annually Statistical Summary)
3. European Gas Pipeline Incidents 1970-92, A Report of the European Gas Pipeline Incident Data Group (EGIG), October 1993
4. US Department of Transportation Statistics, 1970-80
5. PARLOC 96: The update of Loss of Containment Data for Offshore Pipelines 0TH *551* Health & Safety Executive
6. Morgan B. and R.T. Hill (1996) - Current Issues and Questions in Pipeline Safety 1996
7. Morgan B. *(1995),* The Importance of Realistic Representation of Design Features in the Risk Assessment of High Pressure Gas Pipelines. 5th Pipeline Reliability Conference, Houston, Gulf Publishing
8. Fearnehough, G.D. (1978), An approach to Defect Tolerance in Pipelines, IMechE, C97/78
9. Fearnehough, G.D. (1985), The Control of Risk in Gas Transmission Pipelines, IChemE Symposium No. 93
10. Shannon, R.W.E. (1974), The Failure Behaviour of Line Pipe Defects, J. Pressure Vessel and Piping, 2
11. Kiefer, J.F. (1969), Failure Stress Levels of Flaws in Pressurised Cylinders, In: Progress in Flaw Growth and Fracture Toughness Testing, ASTM special publication No. 536, pp461-481
12. Kiefer J.F. (1969), Fourth Symposium on Line Pipe Research, AGA Catalog, No. L30075
13. Morgan B. and Little D. and Beard M (1996) An approach to risk assessment for gasoline pipelines, Pipes & Pipelines International Sep — Oct 1996
14. BS8010 "Code of Practice for Pipelines, Part 2. Pipelines or land design, construction and installation, "Section 2.8 Steel for Oil and Gas, 1992"

Other useful references (not specifically referred to in text)

Bugler, J. (1995), Pipeline risks: the UK regulatory approach. 5th Pipeline Reliability Conference, Houston, Gulf Publishing

California State Fire Marshal (1993), Report of the California State Fire Marshal on Hazardous Liquid Pipeline Risk Assessment, April 1993

Dutch National Environmental Policy Plan – Premises for Risk Management, 1988-89

Improving the Safety of Marine Pipelines : National Research Council, National Assembly Press, 1994

Jones, D.A. and Carter, D.A. (1990), Pipeline Safety Evaluations and their Relevance to Land Use Planning Decisions, Pipeline Management '90, June 1990, London

Jones, D.A. and Gye, T. (1991), Pipelines Protection – How protective measures can influence a risk assessment of pipelines, Pipeline Protection Conference, Cannes 1991

Jones, D.G., Kramer, G.S., Gideon, D.N. and Eiber, R.J. (1986), An Analysis of Reportable Incidents for Natural Gas Transmission and Gathering Lines 1970 through June 1984, Battelle, Columbus, Ohio, Mar 3, 130pp

High Pressure:
Inspection and Maintenance

C571/002/99

Ensuring the integrity of BP Amoco Forties Delta to Forties Charlie crude oil pipeline

S PEET, D G JONES, and **J V PEARCE**
Pipeline Integrity International (PII) Limited, Cramlington, UK
D J WATSON
BP Amoco, Aberdeen, UK

SYNOPSIS

The BP Amoco Forties Delta to Forties Charlie 20 inch diameter crude oil pipeline entered service in 1975 and was inspected by PII in 1998. It was anticipated that the pipeline contained (i) preferential ('slab') internal corrosion at the bottom of the pipe and (ii) internal narrow axial ('groove') corrosion.

Consequently , the inspection involved:-

(i) high resolution magnetic flux leakage (MFL) inspection with additional sensors to detect and size the slab corrosion, and

(ii) a new development, transverse field inspection (TFI), which is specifically designed to detect and size axial groove corrosion.

For the first time the data from the two inspection runs was combined to produce a detailed representation of the corrosion. Deterministic and probabilistic assessments were subsequently conducted using state of the art techniques which allowed future safe operating strategies for the pipeline to be defined.

1. BACKGROUND

The 20 inch diameter, 3.3 km long pipeline which carries stabilised crude oil from Forties Delta to Forties Charlie was laid in 1974 and has operated continuously since 1975. It is predominantly constructed from spiral welded pipe of 12.74 mm wall thickness to API 5L X65 specification. The present requirement is to keep the pipeline operational until 2010.

To date, the only major works carried out have been the installation of emergency shut-down valves, some anode retrofits in 1990 and the Delta topside riser section replacement in 1993 as a result of external and internal corrosion.

Recently the pipeline has operated at 125 bar (MAOP of 129 bar) with a maximum temperature of 55° C , water cut at circa 2% with occasional upset conditions (i.e. higher water cut) and occasionally sand is entrained in the export fluids. A regular programme of operational pigging is now completed every month and corrosion/scale inhibitor is injected.

The pipeline was intelligently pigged in 1991; no significant features were detected. Since then a number of (diver) ultra sonic inspection have been conducted, the most recent was in 1998. This UT inspection revealed areas of significant wall loss in the bottom half of the pipe. There was also the concern that the pipeline contained areas of slab and groove corrosion following the detection of this type of corrosion in a similar pipeline.

On the basis of this knowledge, PII were awarded a risk/reward contract to develop and implement a program of improved inspection techniques through 1998, which could be used to fully evaluate the corrosion mechanisms, determine the integrity status and define the future operating philosophy. This work was part of an increasing focus on integrity management of pipelines which are key to the delivery of future value in the business unit.

In order to ensure that all the different types of corrosion (slab, pitting and groove) in the pipeline were detected, PII devised a program of cleaning, MFL and TFI inspection. Once the inspection results were analysed, a Phase 1 Fitness-for-Purpose was conducted on a deterministic basis to investigate the short term integrity. Subsequently a Phase 2 probabilistic assessment was conducted to investigate the longer term integrity and determine the useful remaining life of the pipeline at a range of conceivable future corrosion growth rates.

2. THE INSPECTION OPERATION

The inspection of steel pipelines using magnetic flux leakage principles is well established. In particular , 'intelligent pigs' have provided reliable in-line inspection for many years. PII's high resolution MFL inspection tool induces a magnetic field in the pipe wall. The presence of corrosion, or indeed any feature that changes the uniformity of the flux path will cause some of the flux to leak out of the pipe wall. This can be readily detected by the circular array of magnetic sensors which are mounted on the inspection tool.

The BP Amoco pipeline was inspected by the PII T3MFL inspection vehicle on 8th August 1998 and by the PII TFI vehicle on 1st September 1998.

The T3MFL vehicle is a development of PII's standard MFL vehicle where the primary sensors have been replaced by Type 3, Hall effect, sensors. These sensors measure the magnetic field strength (e.g. at over 66,000 locations per metre with the 20 inch T3 MFL) at the internal surface of the pipe.

The new primary sensors on the T3MFL vehicle (Figure 1) provide:

- Improved defect detection at very low vehicle speeds.
- The ability to derive the local wall thickness value in the vicinity of each defect.
- The ability to derive the local wall thickness value at each point along the pipeline and
- The ability to detect slab corrosion within the pipeline.

The TFI vehicle (Figure 2) is a new magnetic flux inspection vehicle which is designed to provide improved detection and sizing of narrow, axial oriented, groove-like defects within the pipeline. The TFI's magnetic field is applied (transverse) around the pipe circumference to allow axial defects to be detected by the sensors. The development of the TFI vehicle is fully detailed in Reference (1).

The T3 MFL and TFI inspections detected 5437 metal loss features :-

- the majority were characteristic of internal corrosion at the bottom of the pipe and along the complete length of the pipeline.
- there were over 20 areas of slab corrosion, and.
- 58 axial grooves with depths > 30% wt.

A comparison of the T3 MFL and TFI inspection data from one area of corrosion is shown in Figure 3.

3. PHASE 1 FITNESS-FOR-PURPOSE ASSESSMENT

BP Amoco required confirmation (or otherwise) that the pipeline could safely operate at 115 bar in the short-term. Corrosion is considered to be safe if it is tolerable at a pressure equivalent to 1.39 x 115 bar (1.39 is the generally accepted safety factor implicit in ANSI/ASME B31.G(2)).

A preliminary analysis of the inspection data concluded that if the reported corrosion was assessed according to the requirements of ANSI/ASME B31.G(2), (recognised as being over-conservative), immediate pipeline repairs or de-rating was required. BP Amoco therefore requested the PII Fitness-For-Purpose Group (i) to analyse the inspection data (both TFI and MFL) using alternative approaches to B31.G and (ii) to determine the failure pressure and safe working pressure of the most severe areas of corrosion.

4.1 Linepipe Corrosion Group Sponsored Project

The Linepipe Corrosion Group Sponsored Project (LCGSP)(3) conducted by BG Technology was completed in 1997 and resulted in the production of a set of guidelines which included the assessment of groove corrosion and new rules for defining defect interaction. The findings of the project have been combined with the findings of a parallel project run by DNV. A DNV Recommended Practice RP-F101 on Corroded Pipelines was subsequently issued in April 1999 (4). BP Amoco was a sponsor of the LCGSP and therefore requested that where appropriate the new methods should be used.

4.2 Assessment of Corrosion

4.2.1 Groove Corrosion

The approach developed by the LCGSP (3) is the best available method for assessing the significance of groove corrosion. Figure 4 shows that grooves of length 800mm and maximum depth 46% wall thickness (wt) are tolerable at 115 bar (including an appropriate safety factor) and it was confirmed that infinitely long grooves of 42% wt are tolerable at 115 bar.

All the reported groove corrosion has depths less than 35% wt and is therefore tolerable at pressures in excess of 115 bar.

4.2.2 Complex Corrosion

The RSTRENG (Remaining Strength) approach(5) was developed to remove some of the conservatism in the original B31.G approach. RSTRENG is now accepted world-wide for assessing the significance of complex areas of corrosion in pipelines and is included in the U.S. Department of Transportation (DoT) Pipeline Safety Regulations.

To conduct a full RSTRENG assessment, detailed measurements of the corrosion area are required. For an area of complex corrosion the PII inspection 'boxes' individual areas of corrosion and subsequently 'clusters' them according to recognised interaction criteria. This data can then be processed using PII's LAPA (Length Adaptive Pressure Assessment) software which is designed to conduct an RSTRENG assessment on a cluster (Figure 5).

4.2.3 Interaction Considerations

Defects interact if the spacing between them is such that the failure pressure is lower than the failure pressure of any of the individual defects. The interaction of corrosion is a current area of research.. BP Amoco requested that the interaction rule recommended as part of the LCGSP should be applied to the inspection data. On this basis corrosion that is separated by less than 168 mm was conservatively assumed to interact.

4. FINDINGS OF PRELIMINARY ASSESSMENT OF COMPLEX AREAS

The PII inspection data was re-clustered according to the interaction rule of 168 mm. Consequently the number of clusters reduced from the 5437 originally reported to 2810.

The data was then processed using the PII LAPA and the failure pressure and safe working pressure* of the 2810 features was determined. The local wall thickness data (detected by the T3 sensors on the MFL inspection vehicle) formed a vital input to the assessment. In addition the calculations took into account the combined tolerance effect on both reported defect depth and local wall thickness measurement.

The analysis showed that for the most conservative case (maximum tolerances applied) the failure pressure of the most severe defect in the pipeline was 162 bar equating to a safe working pressure of 117 bar with the application of the 1.39 safety factor.

Deterministic calculations were then conducted to investigate the effect of future corrosion growth (on both grooves and complex corrosion) on the failure pressure. Estimated corrosion growth rates of (0.25, 0.5 and 1.0 mm.year) were applied to the defects to predict when they would reach a size would fall below the predicted surge pressure (Figure 6).

On this above basis BP Amoco's requirement for operation of the pipeline at 115 bar for a 6 month period was justified..

5. PHASE 2 PROBABILISTIC ASSESSEMENT

The Phase 1 assessment confirmed the integrity of the pipeline for operation at 115 bar in the short term. However, the Phase 1 assessment was a deterministic assessment based on lower bound data e.g. minimum wall thickness, maximum corrosion rates, minimum material properties. A probabilistic assessment was therefore conducted to demonstrate the inherent conservatism in the deterministic approach and to determine the probability of failure with time of the pipeline.

Probabilistic methods can be used :-

i) to model the random behaviour of corrosion growth,

ii) to describe the variations associated with the pipe geometry and material properties and

iii) to quantify the probability of failure with time.

Probabilistic assessments have been previously applied and accepted in the nuclear and offshore structures industry and have been used in the limit state design of pipelines. PII have successfully applied probabilistic methods to corroding pipelines since 1993.

It is conventional in a probabilistic assessment not to include any inherent safety factor in the input data. The factor of safety is incorporated by defining a target (acceptable) probability of failure. Currently probabilities of 10^{-4}-10^{-5} are generally utilised for offshore corroding pipelines/risers (4). Target probabilities ranging from 10^{-6} to 10^{-3} are utilised for other engineering structures (6).

---

* Safe Working Pressure = Failure Pressure / 1.39

As only one set of inspection data was available for the pipeline BP Amoco provided estimates of the minimum and maximum conceivable corrosion rates for the future which were based on UT, coupon and inhibitor efficiency data. Corrosion growth rate distributions were then estimated from this data.

Probabilistic calculations were conducted assuming future operating pressures of 129 bar, 65 bar and 29 bar, surge pressures of 17 bar and a range of future conceivable corrosion growth rates (Figures 7-8). e.g. at a target probability of failure of 10^{-4} and applying the low corrosion growth rate the remaining life of the pipeline at 29 bar is 4 years. It was concluded that future operation at 129 bar was not acceptable and that the probability of failure was independent of pressure up to 100 bar.

In addition the feasibility of safely operating the pipeline for at least another 15 years was investigated. The maximum allowable corrosion growth rates were determined at a range of operating pressures which would allow BP Amoco to achieve this target.

Figure 9 shows this relationship at probabilities of failure of 10^{-4} and 10^{-5}, a range of corrosion rates corrosion growth rates and operating pressure of 115 and 83 bar. This information provides BP Amoco with the technical basis for defining future safe operating strategies for the pipeline.

6. CONCLUDING REMARKS

A strategy was devised to determine the integrity of the BP Amoco Forties Delta to Forties Charlie pipeline. By combining the inspection data from both T3MFL and TFI inspections a detailed representation of the significant areas of internal corrosion (slab and axial grooves) was obtained. A staged Fitness-For-Purpose assessment using PII's LAPA software confirmed the integrity of the pipeline at 115 bar in the short term, and a subsequent probabilistic assessment provided BP Amoco with the technical basis for defining future safe operating strategies at a range of pressures and future conceivable corrosion growth rates.

REFERENCES

(1) P.Mundell, K.Grimes, " A new breed of intelligent pig for the detection of defects in the long seam weld of steel pipelines". INSIGHT, Vol 41, No2, February 1999.

(2) Anon, "Manual for Determining the Remaining Strength of Corroded pipeline", The American Society of Mechanical Engineers, ANSI.ASME B31.G 1984.

(3) A.D.Batte, B.Fu , "Assessing the Remaining Strength of Corroded Pipelines", IGE Annual Conference, Birmingham, 19-20 June 1997.

(4) Anon, " DNV Recommended Practice RP-F101, Corroded Pipelines", April 1999.

(5) J..F. Keifner, P.H.Veith, "A Modified Criterion for Evaluation the Remaining Strength of Corroded Pipe, Battelle Final Report PR 3-805 to PRC of American Gas Association , 22 December 1989.

(6) T.Sotberg, R.Bruschi, K.Mork, "The SUPERB Project: A new safety philosophy for submarine pipeline design", OMAE, Florence, June 1996.

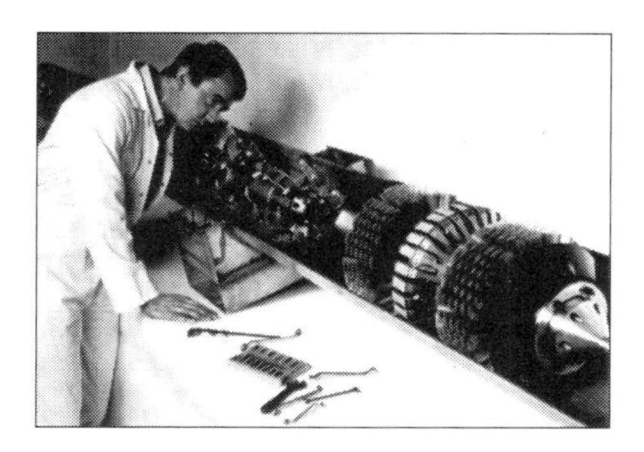

Figure 1 - MFL Vehicle

Figure 2 - TFI Vehicle

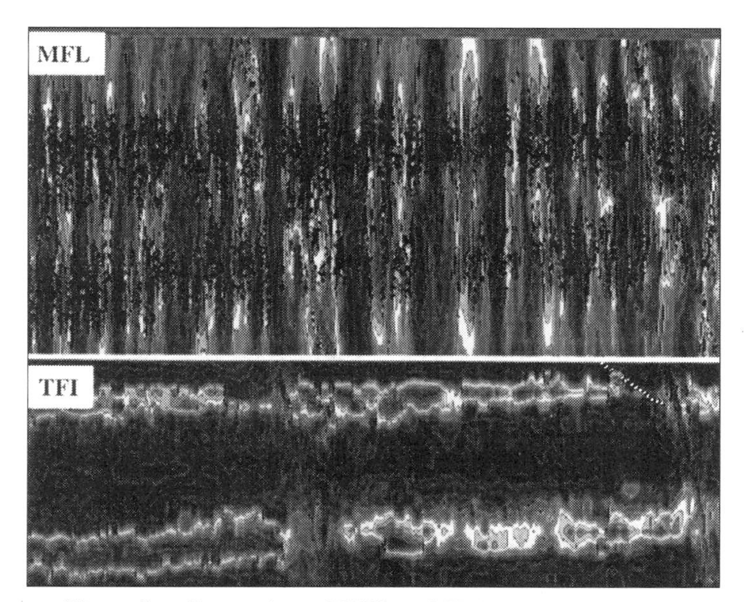

Figure 3 - Comparison of MFL and TFI Inspection Data

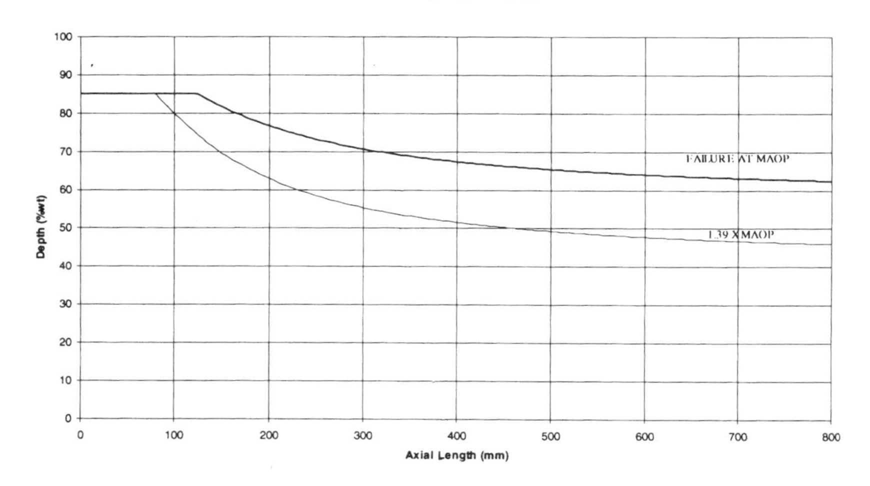

Figure 4 - Size of Groove Corrosion Tolerable at 115 bar

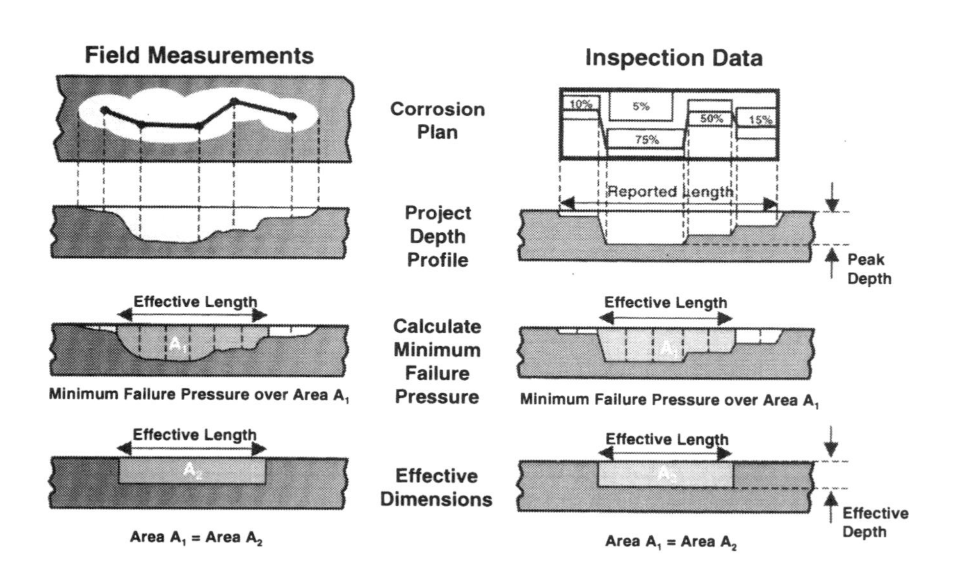

Figure 5 - LAPA = Detailed RSTRENG

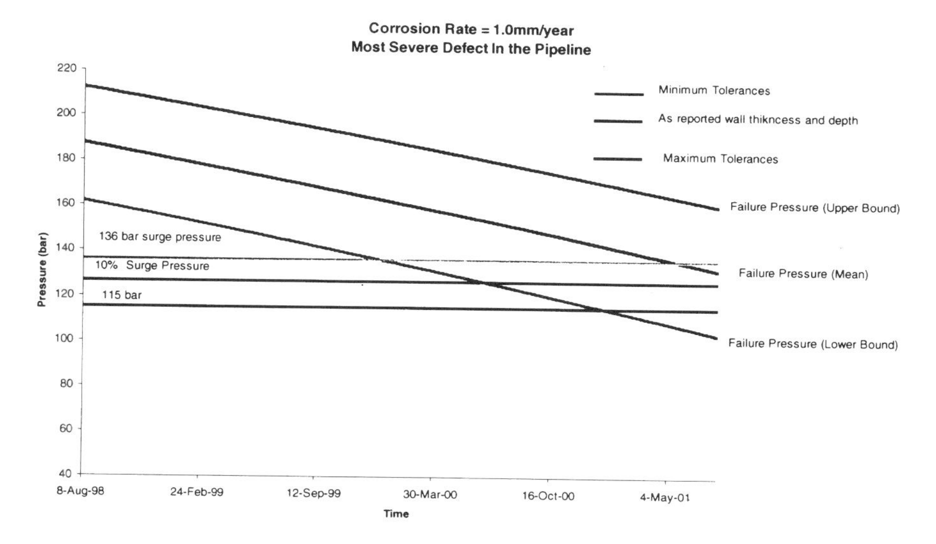

Figure 6 - Reduction in Failure Pressure with a future corrosion rate of 1.00 mm/year

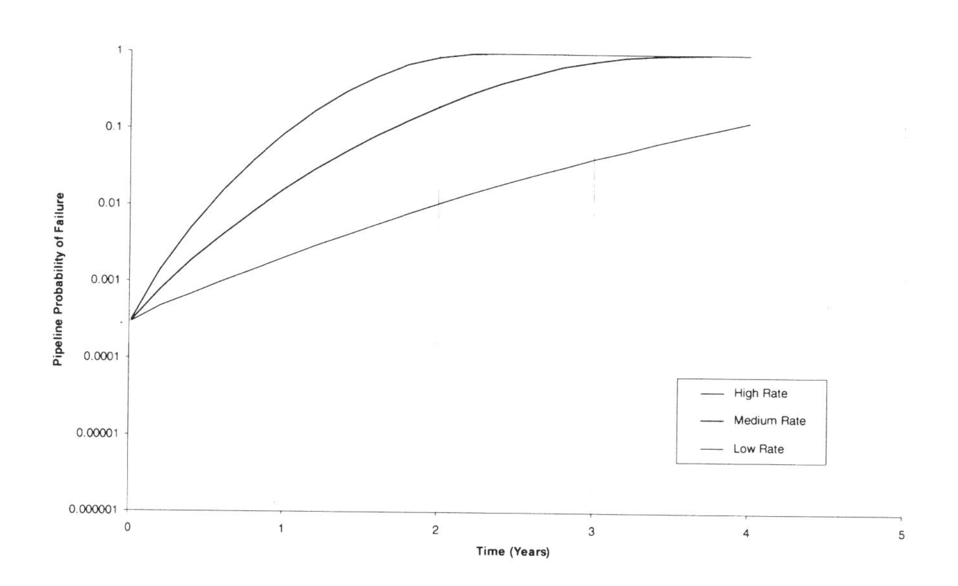

Figure 7 - Pipeline Probability of Failure for Operation at 129 bar

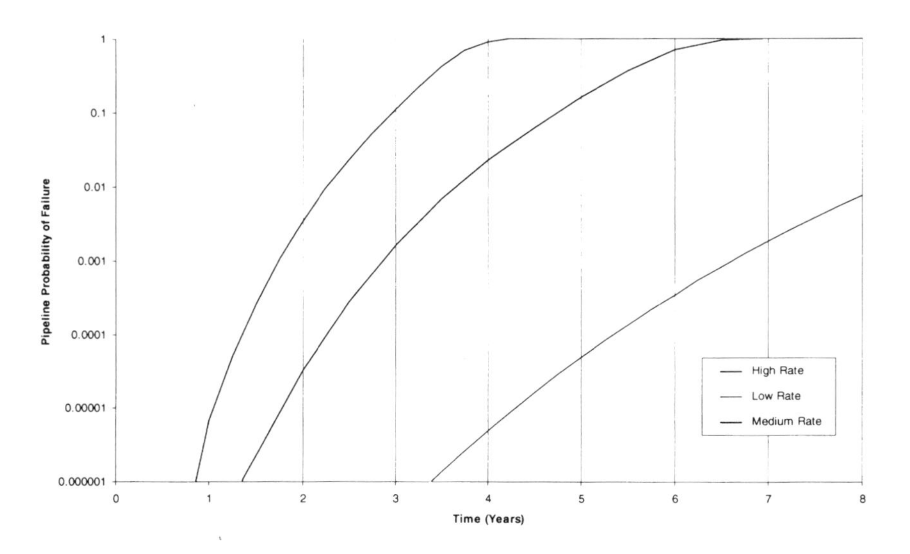

Figure 8 - Pipeline Probability of Failure for Operation at 65 or 29 bar

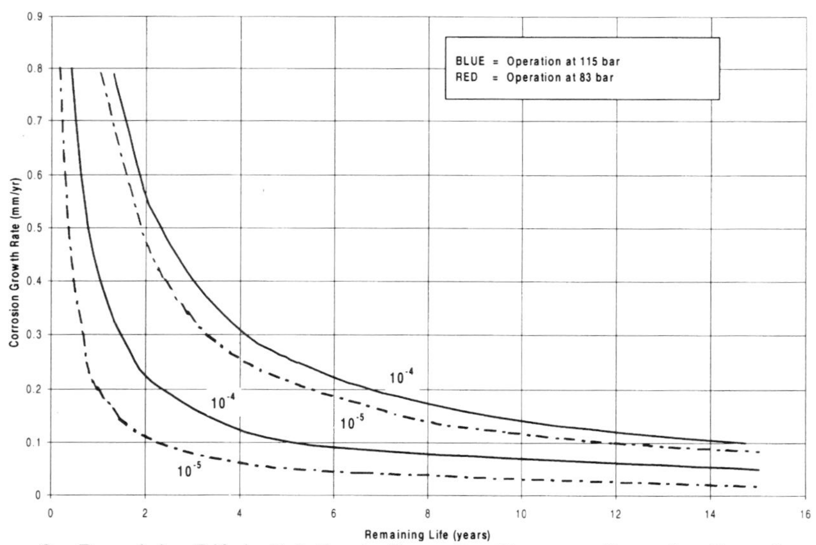

Figure 9 - Remaining Life in Relation to Operating Pressure, Corrosion Growth rate and Probability of Failure

C571/007/99

Assessment of corroded pipework

R GUNN and **G EDWARDS**
Plant Integrity Limited, Cambridge, UK

Abstract

In recent years, new NDE techniques and national standards have been developed for assessment of corroded ferritic steel pipework. These new tools will assist engineers to make financial decisions on whether to leave alone and continue to monitor, renovate or replace.

1 INTRODUCTION

Plant Integrity Limited (P*i*) was set up as a wholly owned TWI (formerly *The Welding Institute*) company in 1997. Part of P*i*'s mission is to add value to its client's operations through the application of innovative inspection and asset management techniques. The following paper describes the latest non-destructive evaluation (NDE) techniques and national standards that have been developed specifically to evaluate process pipework. This approach for pipework is part of the ASPIRE® initiative (Assessment Strategies for Plant Inspection & Repair), whereby the latest concepts of *risk based inspection* (RBI), *fitness for purpose* (FFP) etc are brought together to improve the cost-effectiveness of inspection.

All chemical and petrochemical plants are subject to metal loss with corrosion as the primary cause. Corrosion can take many forms, attacking components from both the inside and outside. During the design stage, the corrosion of ferritic steel components is considered and, usually, corrosion allowances are added to the thickness required to withstand operating pressures etc. Nevertheless, corrosion rates are difficult to estimate at the design stage and often operating plants contain thinned components beyond the corrosion allowance.

Historically, plant inspection was driven by statutory codes and manufacturers' recommendations. As experience has grown, more customised inspection plans have developed with increased inspection of some components, while inspection regimes for other components have decreased, to match the practical situation. This approach is being formalised under the heading of *Risk Based Inspection* (RBI) by such bodies as the American Petroleum Institute, e.g. API document 581 (1), where the *likelihood* and *consequence* of failure for each component is systematically assessed, taking into account safety and economic issues. Vessels are often assigned as high *risk*, due to high replacement costs and their containment of relatively large product volumes. However, plant failure surveys have revealed that piping systems are by far the most likely equipment to cause financial loss (2) by virtue of the significant lengths involved and reduced inspection regimes. In fact, pipework inspection is sometimes considered impractical or prohibitively expensive, due to poor access and the inherent slow rate of conventional NDE methods.

In recent years, several new tools have been developed that can help engineers to assess corroded pipework (see Figure 1). First, new NDE methods have become available that are able to rapidly sort pipework and identify badly corroded areas. Second, corrosion mapping techniques are established that can measure the extent of damage and act as an input to engineering critical assessment (ECA). Third, new BSI, API and DNV documents are to be published in 1999 that include methods for undertaking fitness for purpose (FFP) assessments of corroded pipework based on fracture mechanics principles.

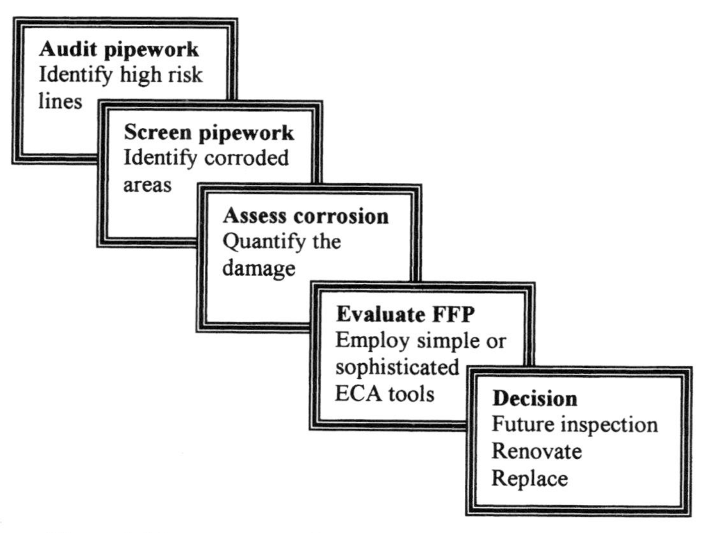

Figure 1: The steps towards an engineering decision

2 FINDING CORROSION

External corrosion is found most easily in ferritic steel pipework through the visible presence of associated corrosion products, that is to say rust. Visual inspection has the advantage of both being a global method for rapid inspection and a local method for examining isolated areas of corrosion. Where there is internal access, boroscopes can inspect a few metres, while CCTV cameras can be used to examine up to 100 meters of pipe, limited only by the length of the fibre optic cable employed.

However, visual inspection is not practicable in many instances, because the surface is not visible. Moreover, gathering quantitative information from visual inspection can be an awkward process. Gaining access to measure the area of corrosion with a rule and the depth of corrosion with a profile gauge is sometimes impossible.

As an alternative, NDE techniques are available for inspecting corrosion in ferritic steel pipework. These are based on the following techniques: magnetic flux leakage (MFL), radiographic testing (RT) and ultrasonic testing (UT). Some are restricted to detecting corrosion on the external pipe surface only and others to corrosion on the internal surface. Many are capable of detecting corrosion on both surfaces, summarised in Table 1 below.

MFL relies on the magnetic flux leakage created around surface imperfections being detected by flux seeking coils. Pipelines are widely inspected from the inside using MFL 'pigs' and a modified technique is widely used as tank floor scanners. The recently introduced *Pipescan* system for process pipework is scanned along the outer pipe surface to detect metal loss.

The use of RT is diminishing, due to the safety issues associated with ionising radiation. High energy gamma radiation is used to penetrate the pipe wall to give an image of the wall thinning on film. Low energy, but high output X-radiation is used with real-time imaging screens to view corrosion on the outer pipe surface under insulation and during operation.

In UT, the most common technique employs digital ultrasonic thickness gauges (DUTG). By timing the flight of pulses of high frequency sound from a probe held on the pipe surface, the ultrasonic thickness gauge is able to measure the pipe wall thickness with an accuracy better than 0.25mm. The *A-scan* display of the reflected pulses on the screen of a conventional ultrasonic flaw detector is still preferred however, since the echo-dynamic patterns can provide important information about the corroded surface, at temperatures up to 400°C. These measurements can be taken at higher temperature, using electro-magnetic acoustic transducers (EMATs).

Table 1: Comparison of corrosion detection techniques for pipework

Technique	External	Internal	Advantages	Disadvantages
Eyeball	Yes, given insulation removal	No	Rapid	Qualitative. Limited to accessible surfaces
MFL, eg. *Pipescan*	No	Yes, given insulation removal	Minimal surface preparation	Detects volume loss not minimum wall thickness. Discrete inspection areas
Gamma RT	Yes	Yes	Wall loss image. Insulation can be left in-situ	Ionising radiation. Discrete inspection areas
X-ray fluoroscopy	Yes	No	Detects corrosion under insulation	Ionising radiation. Tangential image only
Conventional UT flaw detector (A-scan)	No	Yes, given insulation removal	Surface can be scanned for minimum thickness	Requires smooth external surface finish and skilled interpretation
DUTG	No	Yes, given insulation removal	Direct thickness reading	Minimum thickness can be missed
Teletest®	Yes, including buried sleeved pipe and under insulation. Range 2 to 36″ OD	Yes	Rapid with no preparation. Non-intrusive. Non-invasive.	Requires insulation window.
Chime	Yes	Yes	Full circumference from small area access	Limited range and requires insulation window.
Thermograph y	Yes	No	Rapid with no preparation	Detects wet insulation not corrosion
Acoustic emission	Yes	Yes	Limited access required. Scans long lengths	Requires insulation windows, plant shutdown & active corrosion. Interpretation hindered by noise

A recent, highly significant development has been the use of ultrasonic waves, similar to Lamb waves in plates, which are transmitted along the pipe wall from a bracelet of transducers. Known as *Teletest*®, this technique provides 100% coverage and can detect internal and external metal loss under insulated or buried sleeved pipe. Test ranges of up to 30m in each direction from the transducers are regularly achieved and inspection rates of 1km/ 1 mile per day are feasible. Also employing Lamb waves, *Chime* propagates the ultrasound around the pipe circumference from a single transducer placed on the pipe surface. However, to give 100% coverage of a pipe length, a continuous insulation window would need to be removed.

Besides the above NDE methods, techniques for detecting corrosion have also been developed using thermography and acoustic emission. Thermography is used on insulated pipework, where it is able to detect the cold spots associated with damp insulation. This assumes that damp areas have suffered corrosion and that corroded pipe remains in a damp area on the inspection day. Acoustic emission sensors are able to detect ultrasonic emissions from pipe surfaces as they actively corrode. These emissions are present within the general plant noise and so measurements are taken during complete shutdown of connected equipment.

3 CORROSION MAPPING

Given that the remaining wall thickness is beyond the corrosion allowance, it is necessary to produce an accurate wall thickness map, in order to evaluate the fitness for purpose (FFP) of the pipe. From these 'corrosion maps', the area and depth of the corrosion can be measured. If the corrosion is on the outside of the pipe, the surface will have to be mapped laboriously using profile gauges or with setting rubbers. Laser scanning systems are available to automate this process, but are expensive.

If the corrosion is on the inner pipe surface, and its approximate location has been found using one of the techniques described above, then ultrasonic thickness scanning (T-scan) systems can be employed to provide high resolution maps of the corroded surface. From the images, section views can be taken to give corrosion profiles. Alternatively, it is possible, under certain conditions, to derive corrosion profiles by taking tangential shots across the pipe wall with gamma radiation.

4 ASSESSING SIGNIFICANCE

In the current context, *assessment* means undertaking a fitness for purpose (FFP) analysis considering material properties, operating conditions and the geometry of metal loss. There are several approaches to assess the significance of locally thinned areas. Although these approaches have been developed primarily for transmission lines, they can be adapted for process pipework. Relatively smooth corrosion profiles in ductile pipe, where there is little risk of crack-like defects, can be assessed using established loss of area techniques, such as those discussed below. However, particular attention should be paid to the inspection of welds, due to the likelihood of cracks and defects. Where cyclic stresses act on the pipe, crack initiation and growth through fatigue should be considered. Where the corrosion surface is irregular or the absence of cracks cannot be guaranteed, risk of brittle fracture should be assessed using a method such as that described in BS PD 6493: 1991 (3). In addition, for those pipes subject to compressive loads, failure through buckling should be taken into account. In general, the pipe burst pressure will be reduced under the effect of significant longitudinal loads.

For many forms of corrosion, the profile will be comparatively smooth. Here a number of techniques can be used. For a simple quick assessment of locally thinned areas, ASME B31G (4) is adequate, although it may be very conservative. This method requires little more than:

i) the maximum depth and length of the corrosion;
ii) the maximum allowable operating pressure;
iii) the material yield strength.

For a more refined analysis of complex corrosion profiles, Battelle's RSTRENG method (5) may be adopted. This inevitably demands more sophisticated input data, such as that supplied by T-scan, with time and financial implications. However, the benefit of the RSTRENG approach is that it classes more complex corrosion profiles as fit for purpose, than simpler methods will allow.

More precise methods for assessing corrosion damage have been or are currently being developed. API 579 (6) intends to cover a range of corrosion problems with different levels of analysis dependent on the accuracy required of the FFP analysis. However, such techniques are understood to be still under development

Another approach developed from BS PD 6493:1991 into to BS 7910 (7), is currently being coordinated by TWI. It will carry explicit guidance on the consideration of corrosion. Based on validated trials carried out by British Gas (BG), BS 7910 will provide assessment procedures for simple consideration of locally thinned areas, including corrosion in or adjacent to longitudinal and circumferential welds, as well as colonies of interacting corrosion flaws.

Finally, a recommended practice has been produced recently, i.e. DNV RP-F101 (8), as a result of cooperation between BG Technology and DNV. The results of their joint industry projects have been merged, by combining the BG assessment methods with DNV's structural reliability approach.

5 SUMMARY

New screening methods for internal/external corrosion have become available, which combine well with the established methods for detailed corrosion mapping. These NDE methods provide a sound input to new codes that allow corrosion to be assessed in terms of fitness for purpose. As a result, engineers are better equipped to make decisions on the long-term fitness for purpose of pipework and pipelines.

This approach for pipework is part of the ASPIRE® initiative (Assessment Strategies for Plant Inspection & Repair), whereby the latest concepts of RBI, FFP etc are brought together to improve the cost-effectiveness of inspection.

REFERENCES

(1) API 581: 'Base resource document – Risk based inspection', American Petroleum Institute, Preliminary draft, May 1996.
(2) Reynolds J T: 'The application of risk-based inspection methodology in the petroleum and petrochemical industry', PVP-Vol.336, *Structural Integrity, NDE, Risk and Material Performance for Petroleum, Process and Power*, ASME 1996, p125-134.

(3) BS PD6493:1991: 'Guidance on methods for assessing the acceptability of flaws in fusion welded structures', British Standards Institution, 1991.
(4) ASME B31G-99: 'Manual for determining the remaining strength of corroded pipelines', ASME Code for Pressure Piping B31G, The American Society of Mechanical Engineers, 1991.
(5) Battelle: 'Project PR3-805: A modified criterion for evaluating the remaining strength of corroded pipe'. Report to the Pipeline Research Committee of the American Gas Association, Battelle 1989.
(6) API 579: 'Recommended practice for fitness-for-service', American Petroleum Institute, Issue 6, February 15, 1997.
(7) BS 7910:1999: 'Guide on methods for assessing the acceptability of flaws in fusion welded structures', British Standards Institution, 4th draft after public comment, 1999.
(8) DNV RP-F101: 'Recommended practice RP-F101: Corroded Pipelines', Det Norske Veritas, 1999.

C571/008/99

Accounting for uncertainty in the evaluation of pipeline condition based on historical inspection results

C M ROBERTS and **A H S WICKHAM**
Andrew Palmer and Associates, London, UK

SYNOPSIS

During a pipeline's lifetime, inspection technology developments can radically change the preferred survey methods and their accuracy. Reconciling results from disparate inspections while accounting for all uncertainties is a problem.

The paper describes an approach developed to analyse all relevant inspection results holistically. The methodology accounts for all uncertainties associated with pipeline corrosion rate predictions, including fluid corrosivity, inhibition effectiveness, temperature, pressure, flow regime, local pitting and inspection data uncertainties.

The corrosion predictions are continuously updated, based on all available information, including additional general and localised inspection results. The resulting corrosion predictions are therefore always consistent with the known condition of the pipeline.

1 INTRODUCTION

Changes in inspection technology and hence in the preferred methods for and accuracy of surveys may make it difficult to reconcile different sets of inspection results taken over the lifetime of a pipeline. Even where multiple inspections are undertaken using nominally identical tools the error bands associated with those tools impedes the analysis and understanding of the inspection results. The result is that asset managers may be forced to ignore some data and rely only on the findings of later inspections.

To overcome this it is therefore necessary to establish a comprehensive framework within which the results from a number of disparate inspection campaigns can be appropriately reconciled while taking into account all of the uncertainties associated with the condition of a pipeline. A successful framework will allow all of the relevant inspection results to be analysed holistically, thereby ensuring that optimal use is made of all available inspection data.

The paper describes an approach that has been developed to deal with this type of problem, and that has allowed the authors to take into account all the well-known sources of uncertainty associated with corrosion rate predictions in pipelines. These include large scale factors (such as the corrosivity of the transported fluids and the effectiveness of the inhibition regime), medium scale factors (such as temperature, pressure and flow regime) and small scale factors such as local pitting and the uncertainties associated with inspection data.

The methodology was developed to allow continuous updating of corrosion predictions on the basis of all available information such as the findings from additional general and local spot inspections, leak data, and knowledge of no leaks at any point in time. The resulting corrosion predictions are therefore always consistent with the known condition of the pipeline.

This methodology is demonstrated through an example based on a review of the quantitative results of wall thickness measurements obtained through two internal intelligent pig inspection surveys. The first (magnetic flux) intelligent pig inspection survey reported approximately 32000 individual metal loss features having a depth in excess of 10% of wall thickness. The second (ultrasonic) pig inspection was undertaken approximately 2.5 years later. This survey reported approximately 4000 individual internal metal loss features. The significant reduction in the number of individual features reported is caused by three main factors:

1. The difference in accuracy and sensitivity between the two tools used;
2. The additional corrosion resulting in the accretion of many small defects into a smaller number of larger defects with more complex morphologies;
3. Inconsistent defect definition and reporting between the two inspections.

The first point is discussed in more detail in section 3.1 and the second point is addressed in more detail in section 3.2. In order to overcome problem caused by the third point, it was necessary to obtain the basic metal loss data in matrix format. This allowed the re-definition of the defects in a consistent manner.

2 GENERAL APPROACH

Each intelligent pig survey provides a snapshot of the internal condition of a pipeline. The assessment of the condition at the time of survey is a straightforward exercise. The estimation of the useful remnant life of the pipeline is more complex, requiring not only knowledge of the measured defect dimensions at the time of survey but also estimates of the local corrosion rates at the defect sites.

The authors initially considered a number of approaches that could be used to assess the remnant life of a pipeline. The simplest of these would have been to use the wall thicknesses obtained from the two surveys to determine local (pitting) corrosion rates for each identified defect site. Simple extrapolation would then provide an estimate of the wall thicknesses at the time of the analysis or the expected time to through wall penetration or ligament failure. This approach will of course yield only a deterministic assessment of the time to failure (or leak), and therefore would provide minimal input to the decision-making processes relating to the future management of the asset.

Other drawbacks of this approach will be familiar to everyone who has undertaken an analysis of this type. The dates at which individual defects will penetrate through wall are easily defined but confidence limits in those dates are less easy to determine. Data for some defects has to be ignored because the later measurement reports a greater wall thickness than the earlier measurement. Finally, if the analysis is performed some time after the last set of readings was obtained, then individual defects may be predicted to have reached through wall penetration when no leak has been detected.

Perhaps the most significant drawback however is that this type of analysis, although consistent on a defect by defect basis, provides little information with respect to the pipeline as a whole. Asset managers are only concerned with corrosion rates local to individual defects to the extent that they impact on the overall condition of a pipeline. Their real concern is with respect to the remnant life of the pipeline as a whole (or of defined sections of the pipeline). In order to achieve this some assumptions have to be made concerning the statistical independence of the remnant life predictions at each defect.

Assumptions of either complete dependence or of complete mutual independence between defects will allow the analyst to solve the problem. Unfortunately neither assumption can be justified in practice. In reality the real corrosion rates acting within a set of significant defects in a single pipeline will be significantly correlated (all of the defects being subject to attack from the same corrosive fluid and being subject to the same corrosion inhibition regime). The measured corrosion rates will be somewhat less correlated than the real corrosion rates, given that the measured corrosion rates are the real corrosion rates modified by the random measurement errors associated with the pipeline inspection tools.

3 BASIS OF ANALYSIS

3.1 Survey accuracy

The two sets of intelligent pig results were obtained using significantly different instruments. This meant that the accuracy of the measurements obtained during the surveys was also significantly different. As a result of this, a major obstacle in the assessment of the survey data was the reconciliation of the two sets of findings. The accuracy claimed by the operators of the two types of pig is presented in Table 1.

3.2 Reconciliation of survey measurements

The survey estimates of wall thickness are subject to random errors as described above. For some of the defects, the difference between the observed remnant wall thicknesses at the two surveys is small enough that the error bands of the measurements overlap each other to a significant extent. It follows that an assumption of complete independence between the statistical distributions cannot be justified, since this would lead to the possibility that the wall thickness increased between surveys. In order to avoid this outcome it is necessary to reconcile each pair of wall thickness observations by updating the probability distributions for the measurement errors.

A direct qualitative way of visualising this reconciliation is to consider the possibility of the same wall thickness, say 5 mm, being reported in two surveys. Each individual measurement would be modelled by an appropriate distribution having a mean value equal to the observed value and a standard deviation appropriate to the survey methodology. However if the two measurements are considered together it is obvious that, irrespective of the actual numerical values attributed to the measurements, the wall thickness at the time of the first measurement must be equal to or greater than the wall thickness at the time of the second measurement.

The best estimates of the actual wall thicknesses at the times of each of the surveys are modified by knowledge of the observed thicknesses and associated errors at the time of the other survey.

In practice, the effect of the reconciliation of the two sets of survey results leads to an increase in the median corrosion rate over the rate that would be predicted from a deterministic assessment. This increase is greatest where two measurements are close (i.e. where the overlap in the error distributions is greatest). Because the standard deviation of the later ultrasonic measurements is smaller than that of the early magnetic flux measurements, the estimates of the remnant wall thickness at the later survey are relatively unaffected by the reconciliation.

The methodology used to reconcile the survey data utilises a Bayesian approach, which updates the probability distributions used to model each thickness determination. The methodology used is "symmetric in time" in that all defect depth measurements obtained at a specific location are updâted by all other depth measurements made at the same site. The methodology used is general in that it allows an unlimited number of measurements made at a particular location to be reconciled. Implicitly it also allows the concept of "no leak here today" to be automatically incorporated into the analysis.

The methodology explicitly accounts for the correlation between the ongoing pitting corrosion rates at each location within the pipeline; the analysis can therefore be undertaken without making simplistic assumptions about the mutual dependencies between corrosion rates at different locations.

3.2.1 Variations in operating conditions

Throughout a pipeline system, the occurrence of the conditions required for corrosion (water wetting of the pipe wall) is dependent on both the flow velocity and the flow regime. If flow velocities are high enough to prevent water dropout, corrosion will be negligible. If flow is slugging two-phase, corrosion conditions may occur. If the flow rate is negligible, resulting in near stagnant conditions, significant water dropout will occur and corrosion will be expected. Where wetting of the pipewall is expected corrosion inhibition is the only protection against high levels of metal loss.

The analysis software developed included a corrosion rate prediction model, which was used to estimate both the historic cumulative corrosion over a specified time interval and the predicted future corrosion rate under assumed operating conditions. The difference between the predicted average corrosion rate over the time interval between intelligent pig surveys and the predicted future corrosion rate based on assumed future operating conditions was calculated. This was used to modify the predicted failure dates to take account of changes in the corrosivity of the transported fluids.

For the results presented here, no changes in the composition of the fluids or in the operating conditions were applied, however the authors have applied the same methodology to a case involving a significant change in the corrosion inhibition regime.

3.2.2 Availability of corrosion inhibition

An additional random distribution was introduced into the analysis to represent the availability of the corrosion inhibition from the date of the last survey forward into the future. The corrosion inhibition availability must take a value between 0.0 and 1.0. For this reason, it was modelled using a Beta distribution with finite upper and lower tails. For the base case analysis, the Beta distribution parameters were selected to give a mean value of 0.9 and a standard deviation of 0.01. Modelling the inhibition availability as an uncertain quantity increases the variability of the calculated times to failure. The small standard deviation was assumed to avoid an adverse influence on the results of the base case assessments. The potential effects of a higher or lower level of corrosion inhibition availability can be examined by modifying the appropriate parameters and repeating the analysis.

3.3 Determining the remnant life

In order to maximise the information available to the asset manager the distributions of the dates of the first five leaks or ligament failures expected with each pipeline section analysed were determined. This was in addition to the distributions of the equivalent dates for each defect individually.

3.3.1 Data used in analysis

The 30 defects[1] having the smallest remnant wall thicknesses reported in the most recent ultrasonic intelligent pig survey were selected for the analysis. In order to establish the local corrosion rates effective between the dates of the two surveys it was necessary to establish the metal loss reported in the earliest survey at the corresponding location. Establishing this information involved the usual pipe joint by pipe joint comparison of reported lengths to determine the equivalent pipe joint numbering for the two surveys and derive the offset of each reported defect from the start of the pipe joint in which it lies. For the defects examined remnant ligament thicknesses at the most recent survey varied from 7.5mm to 3.2mm (nominal wall thickness was 12.7mm, actual average was 13.3mm).

3.3.2 Through wall corrosion

Table 2 presents the results of this study in respect of through wall corrosion. The analysis results indicate that the median estimate for the date of the first leak from the set of 30 defects considered here is July 2001. The 5% and 95% confidence limits for the date of the first leak are October 2000 and July 2002 respectively. The table also shows the equivalent dates for the predicted occurrence of the second to fifth leaks.

Figure 1 shows the cumulative probability of through thickness corrosion for each of the 30 defects plotted against time. The steepness of the curves is indicative of the rate of corrosion associated with each individual defect. It is clear that there is a significant variability between the defects in the median predictions of the date for through wall penetration (i.e. 2002 to 2021) although the 1998 observed wall thicknesses for all but one of the defects considered, cover a range of only 1.4 mm.

Figure 2 shows the probability of through thickness penetration by one, two, three, four and five of the defects plotted against time. Figure 3 shows the relative frequency for each of the defects of providing the initial through wall penetration. Defects 5 and 13 are dominant and

---

[1] The methodology generalises to any number of defects and the authors have subsequently applied it to the simultaneous analysis of up to 125 defects in a single section of pipeline.

together account for 55% of initial penetrations.

A sensitivity analysis of the results demonstrated that the analysis is relatively insensitive to the inclusion or exclusion of any individual defects. This reflects the fact that within the sample considered, the defects all have approximately the same depth and corrosion rate. It also reflects the relatively large uncertainty in the observed corrosion rate, which is in turn driven by the (relatively) large depth measurement uncertainty associated with MFL inspection tools and by the short time interval between the two surveys.

4 CORROSION ASSESSMENT OF PIPELINE

The whole pipeline was assessed on a section by section basis. The analysis performed was the same as that described above for the first section of the pipeline. The pipeline was sectioned for detailed analysis on the basis of the local variations in metal loss defect frequency.

Within each section the 30 locations having the deepest corrosion were analysed as described previously. The results of the analysis (expressed in terms of the date of through thickness corrosion and structural failure) are given in Table 4.

As expected, when taking structural failure into account the estimated remnant life is less than when considering through thickness penetration only. It should be noted however that there is more uncertainty associated with the ligament failure estimates because they depend not only on the data obtained in the various surveys but also on assumptions about the pressure in the pipeline and the failure stress in the ligaments (1).

5 DECIDING WHAT TO DO NEXT

The results of such a remnant life assessment obviously require a reaction. With a low remnant life it is necessary to make a rapid decision to repair or replace the pipeline. Such a decision can be highly contentious and political with different stakeholders pushing for different outcomes. Any intuitive decision is very difficult to justify.

Modern, risk based decision making techniques, as proposed in (2) and (3), allow alternatives to be considered by accounting for all consequences of failure such as the loss of life, environmental damage or damage to company reputation in terms of a non-dimensional value. This accounts for:

- the decision maker's attitude to risk;
- the value of the consequence in a convenient and meaningful dimension;
- any non-linearity between actual value and the effect of the value on the organisation;
- the relative importance of each factor to the stakeholders.

The use of risk based decision methods can help to work through the controversies described previously. The greatest benefit of the risk based decision methods is the rational and auditable manner in which the decision problem can be fully explored. Significant risks can be evaluated and quantified and a variety of opinions, values and judgements can be accounted for and combined. Finally a preferred course of action can be identified and agreed upon through discussion and consensus building. This process when applied carefully will lead to better considered, justifiable and explicable decisions. More detailed descriptions of the methods available and example applications are presented in (4) and (5).

6 CONCLUSIONS

The paper has described an approach that was developed to deal with the problem of reconciling disparate inspection results. The methodology described takes account of all significant sources of uncertainty associated with corrosion rate predictions in pipelines. Previous methods for predicting pipeline corrosion rates based on historical inspection results and/or theoretical corrosion models have not succeeded in attaining the right balance between statistical independence and correlation of corrosion rate along the length of the pipeline.

The approach presented here has taken account of these conflicting issues in a statistically rigorous manner. The partial correlation of the corrosion rate along the pipeline was accounted for by developing a comprehensive model which considers:-

- macro scale correlations associated with the general corrosivity of the fluids and the corrosion inhibition regime in place in the pipeline;
- medium scale correlations associated with flow regime variations due to changes in temperature and/or pressure and/or bathymetry; and,
- micro scale independence of measurement errors from defect location to defect location.

An assessment of this type, using all available data, results in an increased confidence in the understanding of the current and future state of the pipeline. This understanding directly supports the decision making that then follows concerning monitoring, and repair and/or replacement schedules for the pipeline.

7 REFERENCES

1 Rosenfeld, M. J., Kiefner, J. F.; 'Proposed Fitness-for-Purpose Appendix to the ASME B31.8 Code for Pressure Piping - Section B31.8, Gas Transmission and Distribution Systems', Kiefner and Associates, Worthington, Ohio, Jan. 1995.

2 Keeney, R. L. and Raiffa, H. (1976): "Decisions with Multiple Objectives: Preferences and Value Trade-offs", John Wiley and Sons.

3 Belton, V. (1990): "Multiple Criteria Decision Analysis - Practically the only way to choose", Operational Research Tutorial Papers, Ed. Hendry, L. C. and Eglese, R. W., Operational Research Society.

4 Roberts, C. M., Strutt, J. E. and Shetty, N. K. (1998): "Using Risk Based Decision Analysis to Manage Structural Assets" in Safety and Reliability: Proc. of the European Conference on Safety and Reliability, Esrel '98, Vol. 1, pp181-188, AA Balkema.

5 Roberts, C. M. (1999): "A Review Of Decision Analysis Methods For Application To Risk Based Management Of Structural Assets" Cranfield University School of Management Working Paper No. SWP 8/99.

Table 1: Quantification of uncertainties associated with each PIG type.

PIG type	Reporting threshold for general corrosion	Reporting threshold for pitting corrosion	Percentage of measurements within confidence limit	Confidence limits on measurements	Implied standard deviation
MFL	10% WT	30% WT	80%	± 10% WT	1.038 mm[1]
Ultrasonic	N/A	N/A	95%	±0.5mm[2]	0.255 mm

Note 1: Based on overall average 13.3 mm wall thickness

Note 2: Subject to depth measurement being over flat area with diameter greater than or equal to 20 mm.

Table 2: Predicted dates of through thickness leaks (30 defects)

Probability	First Leak	Second Leak	Third Leak	Fourth Leak	Fifth Leak
5%	Oct-2000	Apr-2001	Aug-2001	Nov-2001	Mar-2002
10%	Dec-2000	May-2001	Sep-2001	Jan-2002	May-2002
25%	Mar-2001	Sep-2001	Jan-2002	May-2002	Sep-2002
50%	Jul-2001	Jan-2002	May-2002	Sep-2002	Jan-2003
75%	Dec-2001	May-2002	Oct-2002	Mar-2003	Jul-2003
90%	Apr-2002	Oct-2002	Mar-2003	Aug-2003	Jan-2004
95%	Jun-2002	Jan-2003	Jun-2003	Nov-2004	May-2004

Table 3: Predicted dates of through thickness leaks

Probability	Ligament Failure / Through Thickness Corrosion	Through Thickness Corrosion Only
5%	September-1999	October-2000
10%	October-1999	December-2000
25%	January-2000	March-2001
50%	March-2000	July-2001
75%	July-2000	December-2001
90%	October-2000	April-2002
95%	January-2001	June-2002

Table 4: Predicted dates of through thickness leaks and structural ligament failure

	Through Thickness Leaks			Structural Ligament Failure		
Section	Median Estimate	5% Confidence	95% Confidence	Median Estimate	5% Confidence	95% Confidence
1	Jul-2001	Oct-2000	Jun-2002	Mar-2000	Sep-1999	Jan-2001
2	Jul-2007	Jul-2005	Mar-2010	Nov-2002	Sep-2001	Apr-2004
3	Jan-2004	Mar-2002	Jun-2006	Aug-2002	Apr-2001	Jan-2004
4	May-2012	Oct-2008	Mar-2017	Mar-2007	Jul-2004	Jun-2011
5	Sep-2003	Jun-2002	Apr-2005	Oct-2001	Jan-2001	Sep-2002
6	Jan-2006	Mar-2004	May-2008	Aug-2002	May-2001	Jan-2004
7	Oct-2003	May-2002	Apr-2005	May-2002	Jan-2001	Mar-2004

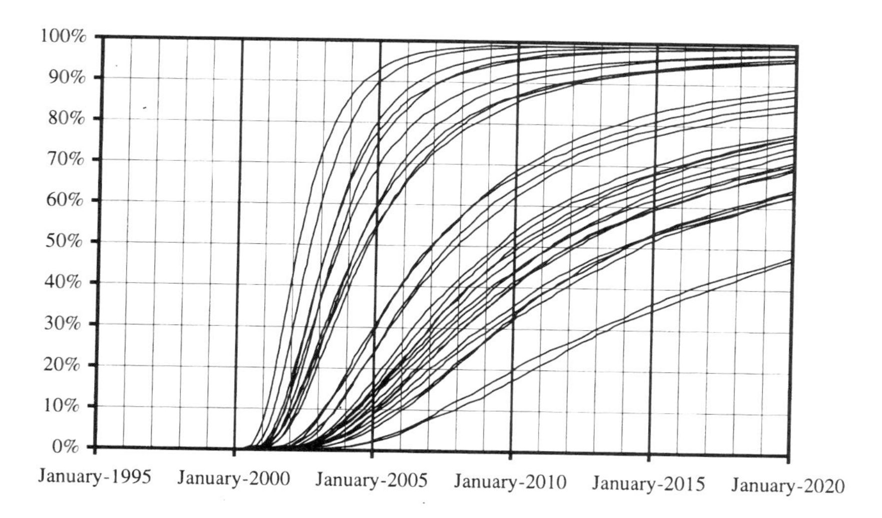

Figure 1: Probability of penetration, individual defects

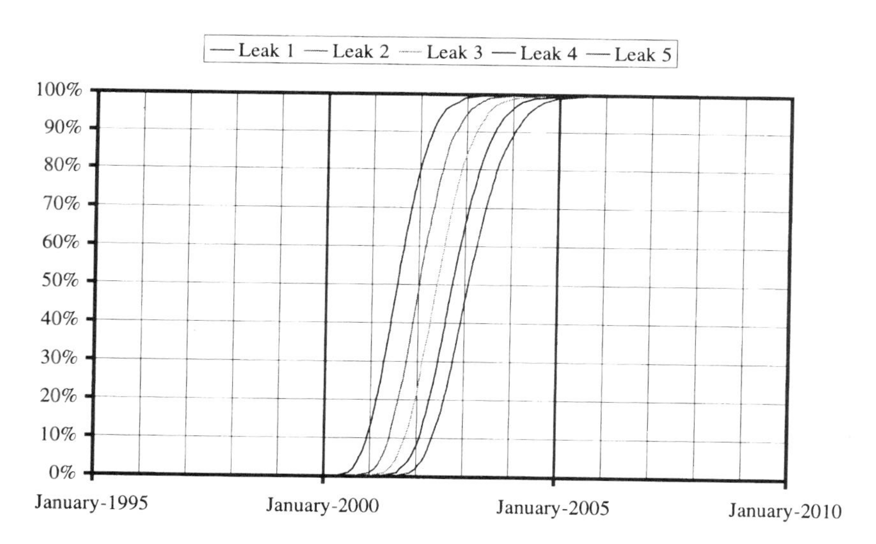

Figure 2: Probability of penetration, all defects

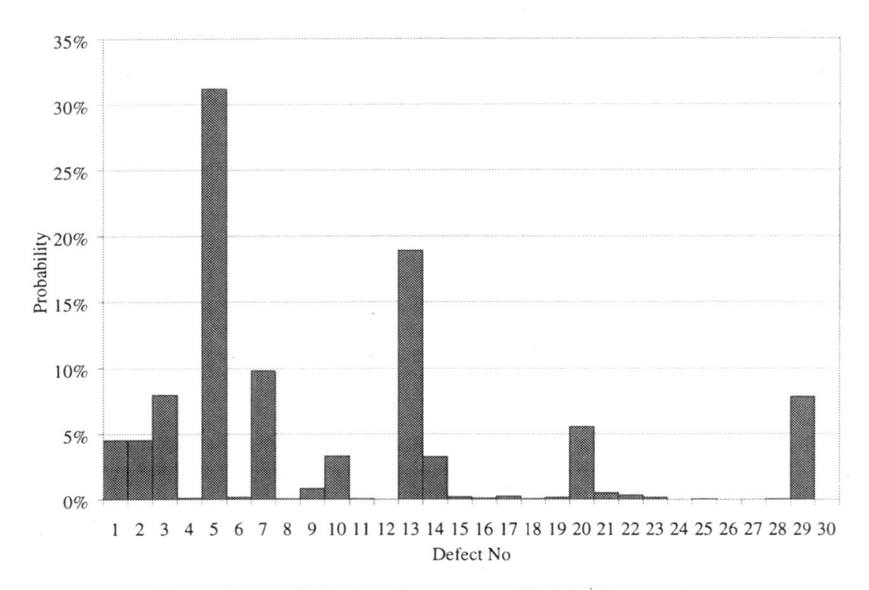

Figure 3: Relative Frequency of Initial Penetration

C571/029/99

Shell Pipeline Risk Band Inspection

H J M JANSEN
Shell International Exploration and Production B. V., Rijswijk, The Netherlands
B F M POTS and **C W M VOERMANS**
Shell International Oil Products B. V., Amsterdam, The Netherlands

ABSTRACT

This article describes the results of application of the Shell in-house developed PIPE-RBI methodology on a number of pilot pipelines. PIPE-RBI has been successfully proven by the pilot cases on a wide variety of pipelines. The examples given by the pilot study clearly indicate the large benefits that can be obtained from the PIPE-RBI approach. Features it provides are:

- Structured process for pipeline integrity management and audit trail;
- Pro-active focus on corrosion control;
- Tool for rehabilitation planning;
- Frequency setting of intelligent pig inspection by Risk Based Inspection;
- Enhanced control of pipeline technical integrity;
- Relatively easy to apply.

Potential cost saving by deferring inspection is shown for an offshore gas pipeline of 1.2 million USD. Another example of an on-shore oil pipeline shows a Net Present Value cost saving of potentially 2 million USD by extending the pipeline's life.

1. INTRODUCTION

Key activities in the integrity management process of pipelines are intelligent pigging inspection planning and integrity status assessment. Considerable cost savings can be realised when utilising Risk Based Inspection (RBI) methodology. For example, RBI techniques generally yield longer inspection intervals compared to time-based inspection; are effective in prioritising pipelines when only limited inspection resources are or can be made available; and can give the confidence for safely postponing repair or replacement of damaged pipelines.

Shell has developed and is implementing a RBI methodology for pipelines. The Shell Risk Based Inspection (S-RBI) approach has been embedded in the Pipeline Integrity Process loop and integrated with tools for corrosion rate assessment, remaining life assessment and rehabilitation planning, see Figure 1. The MS-Access data management system "PIPE-RBI" has been developed by which the S-RBI methodology and support tools can be applied.

The PIPE-RBI methodology focuses on controlling internal corrosion. A project is on-going extending the methodology to external corrosion control. Another project has been initiated to develop methodology for corrosion growth estimation between two intelligent pig inspections. This methodology will be included in the ASSESS-PIPE software, which is Shell proprietary software for assessing intelligent pig data.

In this article the pipeline RBI methodology, as implemented in PIPE-RBI is explained and a number of application examples are described, illustrating the gains that can be made.

METHODOLOGY

2.1 Risk Based Inspection method

The S-RBI method for pipelines is based on the generic S-RBI methodology, which was developed in the Shell in-house Alliance Risk and Reliability Management project with a focus on vessels and pipe-work. The methodology takes account of the risks associated with a pipeline's functional failure from corrosion, where the risk of functional failure is determined by the probability and the consequence of such failure.

For age-related degradation mechanisms, such as corrosion, the inspection interval is based on the remnant life and the Inspection Interval Factor.

Maximum Inspection Interval = Inspection Interval Factor x Remnant Life
The Inspection Interval Factor is a function of criticality level and confidence rating, which are defined as:

Criticality level:

> The pipeline criticality assessment is based on the Risk Assessment Matrix document issued by the Shell Health, Safety and Environment Committee. Criticality is a function of failure consequence and probability. The level of probability or susceptibility to failure for corrosion is determined as the ratio between the actual corrosion rate and the design corrosion rate. Failure consequence rating is assessed via questionnaires.

Confidence rating:

> The Confidence Assessment reflects the confidence in the assessment of the remnant life and depends as such on the accuracy or variance of both the actual corrosion rate and the corrosion allowance or wall thickness redundancy. The confidence rating is determined from a standardised questionnaire, which need to be filled in by the integrity engineer doing the analysis.

2.2 Modules for corrosion risk and damage assessment

S-RBI is only the framework of the methodology. Actual evaluation of the probability of failure and the remaining life hinges on a number of modules, notably those for corrosion degradation and defect assessment.

The corrosion degradation module is called ASSESS-COR and combines the corrosion rate information from intelligent pigging inspection and corrosion prediction model calculations (1), (2). The model calculations are performed utilising the computer program HYDROCOR, taking into account corrosion degradation caused by: CO_2, H_2S, organic acids, O_2, bacteria, or a combination. The HYDROCOR model calculations provide a link between the pipeline operational conditions, the corrosion risk and corrosion control measures. An important element in ASSESS-COR is the way weights are given to the various corrosion rate information sources. For example, when multiple, reliable IP inspections are available, the corrosion rate is mainly determined from inspection; if no inspections are available, the model calculations determine the corrosion rate. When insufficient information is available for a model calculation, use is made of a default corrosion rate, making use of corrosion circuits.

The ASSESS-PIPE module is used for (a) collecting inspection data from Intelligent Pig surveys and individual monitoring points and (b) assessing the mechanical integrity of the pipeline (3). The reported corrosion defects are assessed versus the required maximum allowable operating pressure according to the Shell92 defect assessment code (4). ASSESS-PIPE is now an integrated part of the RBI process. It is used to calculate the minimum corrosion tolerance in the pipeline on basis of the corrosion defects as found by the inspection. An additional functionality is the sensitivity analysis for repair and replacement planning. This will help the asset manager to evaluate several maintenance scenarios and therefore minimising maintenance and operational costs, e.g. repair versus required inspections in due time.

2.3 Data management

By investing in proper data management, the integrity management process can be automated resulting in the large potential benefits of the PIPE-RBI approach being realised. A good example of linking the various data sources is given by the Shell EXPRO concept. The attached Pipeline Integrity Management System diagram in Figure 2 illustrates how the pipeline asset register, physical data and reporting database (Common Pipeline Database CPD) relates to other key EXPRO systems.

3. PILOT EXAMPLES

3.1 Off-shore 36-inch gas pipeline

The first example is an offshore 36-inch gas export pipeline of 450km length, which was due for intelligent pig re-inspection in 2000. The main integrity threat is internal CO_2 corrosion resulting from periods of wet operation. The pipeline started operation in 1982 and was inspected by intelligent pig in 1991 when no corrosion was found.

The corrosion history in the pipeline has been built-up by the ASSESS-COR analysis as indicated in Figure 3. Wet operation can occur from inadequate dew point control and operational upsets (leading to free water entering the pipeline). No information was available on operational upsets leading to free water entering the pipeline; 1% wet operation from upset conditions was conservatively assumed. No data on dewpoint control are available before 1994, hence a conservative 98% availability of the dewpoint control system is assumed between 1982 and 1994. Dewpoint control has been monitored since 1994. The HYDROCOR results in Figure 3 show short periods of dewpoint off-spec conditions in April and August 1995. Since 1996, dewpoint has been on target. As a result of monitoring the main process parameters, it can be safely assumed that the internal corrosion is controlled to below 0.05 mm/y.

The ASSESS-COR analysis shows that localised corrosion monitoring probes are unsuitable to quantify internal corrosion rates. Usually the corrosion monitoring probes give far lower corrosion rates than the worst corrosion rate occurring in the pipeline. This is also illustrated by other pilot examples.

The basis for the remnant life assessment is the estimated Remaining Wall Thickness (RWT) over the pipeline's life, see Figure 4. The Minimum Allowable Thickness is based on the Shell92 defect assessment method whereby a corrosion geometry factor is taken into account. For the gas pipeline, the end of remnant life is calculated to be in 2036, based on the conservative corrosion assessment.

The criticality and confidence assessments result in an Inspection Interval Factor of 0.4. The Inspection Due Date is 1991 + 0.4 * (2036 – 1991) = 2009. Future intelligent pig runs may be further deferred if operators monitor and record the annualised % wet operation, including upset conditions leading to free water entering the pipeline.

The functionality of building up the corrosion history has been used effectively in determining the inspection due date. Application of PIPE-RBI allows extension of the inspection interval from 9 to 18 years. The OPEX cost saving is estimated to be 450 km * 2500 USD/km + 50,000 USD ≈ 1.2 million USD.

3.2 Offshore 18-inch gas/condensate pipeline

The second example is an offshore 18-inch gas/condensate pipeline of 16km length in the North Sea. The pipeline has been in operation since 1986 and was inspected in 1997 by an intelligent pig where no corrosion defects were found.

For this pipeline, uncertainty existed on % wet operation, because the operational upsets and dewpoint control were not adequately monitored. Without making use of the PIPE-RBI analysis, the corrosion engineer assumed 5% wet operation on basis of some known operational upsets and a period of wet operation equalling the period between de-watering pig runs.

A sensitivity analysis was made on % wet operation making use of the ASSESS-COR methodology. On basis of the 1997 intelligent pig inspection, it was concluded that the average % wet operation was below 2%, see Figure 5. As such, a worst case corrosion rate of 0.09 mm/y was assumed. The resulting end of remnant life and Inspection Due Date are respectively 2028 and 2009.

The operating company had been using a fixed inspection interval of 5 years. Application of PIPE-RBI allows extension of this interval to 12 years, saving over 50% on life-cycle inspection costs. Cost of an intelligent pig inspection for this pipeline is estimated to be about 130,000 US$.

Despite the reduction in intelligent pig inspection, it is argued that the technical integrity of this pipeline is better maintained than previously. As a result of the PIPE-RBI analysis, the main process parameters controlling wet operation are now monitored by which the % wet operation will be reduced, hence pro-actively preventing corrosion to occur.

3.3 Onshore 20-inch oil pipeline

The third example is an onshore 20-inch oil pipeline of 51km length in the Middle East. Intelligent pig inspections of 1994 and 1997 indicate significant corrosion damage in the pipeline. Since 1994, the pipeline has been chemically inhibited by which the internal corrosion is controlled to about 0.1 mm/y.

The remaining corrosion tolerance for each defect has been determined via the ASSESS-PIPE analysis as illustrated in Figure 6. The corrosion tolerance is the amount of corrosion that can be tolerated for the defect still complying with the Shell92 defect acceptance criteria. The ASSESS-PIPE methodology has a major impact on the prediction of estimated remnant life and rehabilitation planning. Repairs can extend the remnant life in a structured manner, making use of the corrosion tolerance concept, see Figures 7 and 8.

The PIPE-RBI methodology enables scenario planning to enable this pipeline to be operated in the most cost-effective way. The effects of de-rating, repairs and improved corrosion control can then be calculated with respect to the remnant life. A Net Present Value calculation has been carried out for the original rehabilitation plan and a revised plan in accordance to the PIPE-RBI analysis, the results of which are given in Table 1. The NPV cost saving between the two scenarios is in excess of 2 million USD, mainly as a result of pipeline life extension.

PIPE-RBI integrated with ASSESS-PIPE, enables life extension by several years and as such further sweating of the asset in a controlled manner. Added to the other preservation efforts, such as an enhanced pigging, inhibition and repair programme, this contributes to a significant NPV cost saving on replacement CAPEX.

4. CONCLUSIONS

The integration of risk based planning, corrosion modelling, intelligent pig inspection and technical integrity assessment provides a closed loop of plan, schedule, execute, analyse and improve to manage pipeline integrity with respect to the internal corrosion threat. The MS-Access data management system "PIPE-RBI" has been developed by which the S-RBI methodology can be applied. The strength of the PIPE-RBI methodology is its modular set-up and linkage to the ASSESS-COR and ASSESS-PIPE tools.

The approach has been successfully tested on a number of pilot pipelines from six Shell Operating Units around the world. On basis of the pilot study, it is expected that the intelligent pigging programme can be reduced by 25%. The additional advantages of the approach are:

- An enhanced control of pipeline integrity;
- A structured process for rehabilitation planning;
- A reduction on CAPEX for the pipeline replacement programme.

REFERENCES

1. C. de Waard, U. Lotz and D.E. Milliams, "Predictive model for CO2 corrosion engineering in wet natural gas pipelines", CORROSION/91, paper 577, 1991.

2. B.F.M. Pots, "Mechanistic models for the prediction of CO2 corrosion rates under multi-phase flow conditions", NACE CORROSION/95, paper 137, 1995.

3. W.R. Vranckx and R. Schofield, "Pipeline Integrity Management in Petroleum Development Oman; Developments in intelligent pigging surveys, defect assessment and pipeline repair techniques", Pipelines '97, 23-24 February 1997, Dubai UAE.

4. D. Ritchie and S. Last, "Burst criteria of corroded pipeline, defect acceptance criteria", Shell external report AMER.96.010, June 1996.

Year	Original plan			Revised plan by PIPE-RBI		
	Activities	**Real Term in USD**	**M.O.D. in USD** (3% inflation)	**Activities**	**Real Term in USD**	**M.O.D. in USD** (3% inflation)
1999				2 repairs	20,000	22,510
2000	IP	127,500	147,807	2 repairs	20,000	23,185
2001	8 repairs	80,000	95,524	IP	127,500	152,242
2002	8 repairs	80,000	98,390			
2003	8 repairs	80,000	101,342			
2004	8 repairs	80,000	104,382	IP	127,500	166,359
2005	Replace	20,000,000	26,878,328			
2006						
2007						
2008				Replace	20,000,000	29,370,674
NPV (8% discount):			**16,093,000**			**13,870,000**

Table 1: Example of potential Nett Present Value cost savings by application of PIPE-RBI for onshore 20-inch oil pipeline.
(Replacement cost: 20 million USD; repair cost: 10,000 USD: IP cost: 127,500 USD)

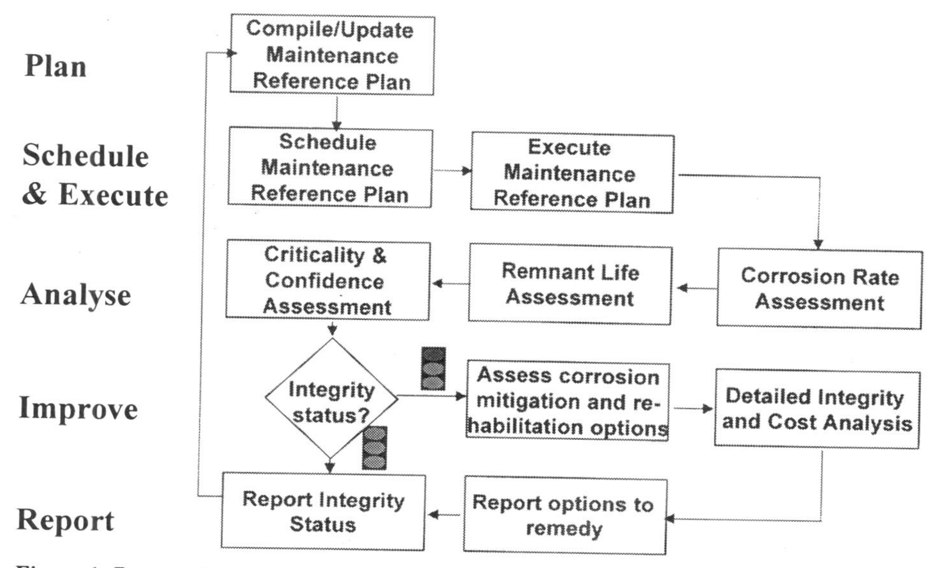

Figure 1: Process for Integrity Management and Risk Based Inspection. (Maintenance Reference Plan includes a list of all maintenance, inspection and corrosion control tasks)

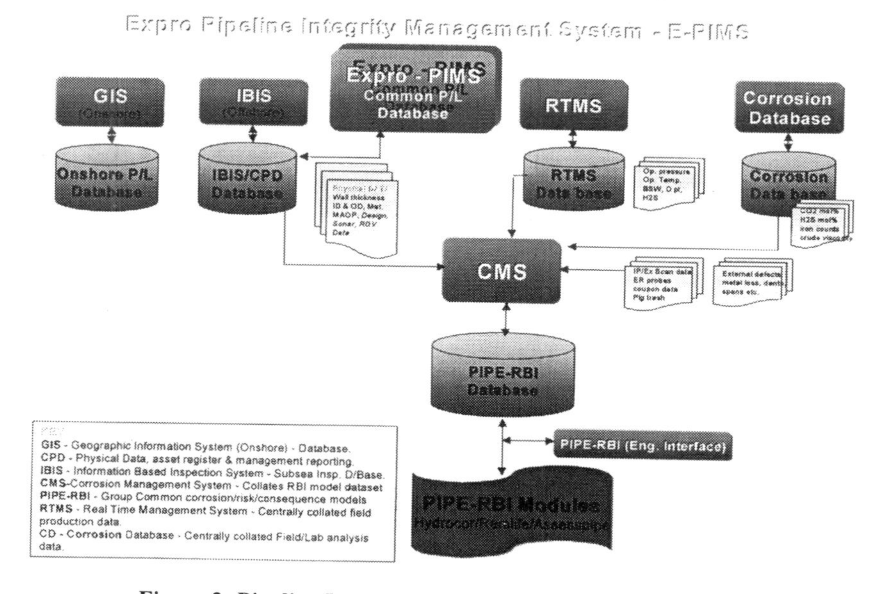

Figure 2: Pipeline Integrity data management of Shell EXPRO

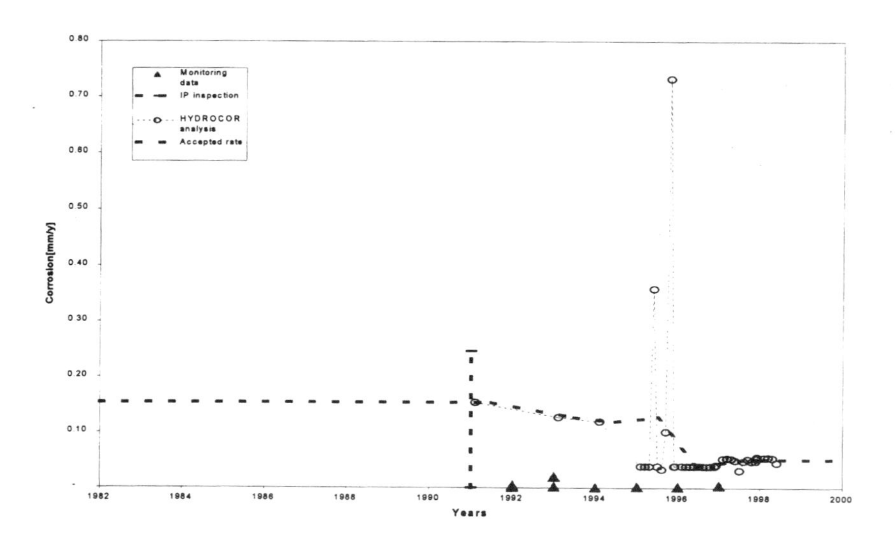

Figure 3: ASSESS-COR analysis for offshore 36-inch gas pipeline. HYDROCOR analysis indicates a future corrosion rate of 0.05 mm/y.

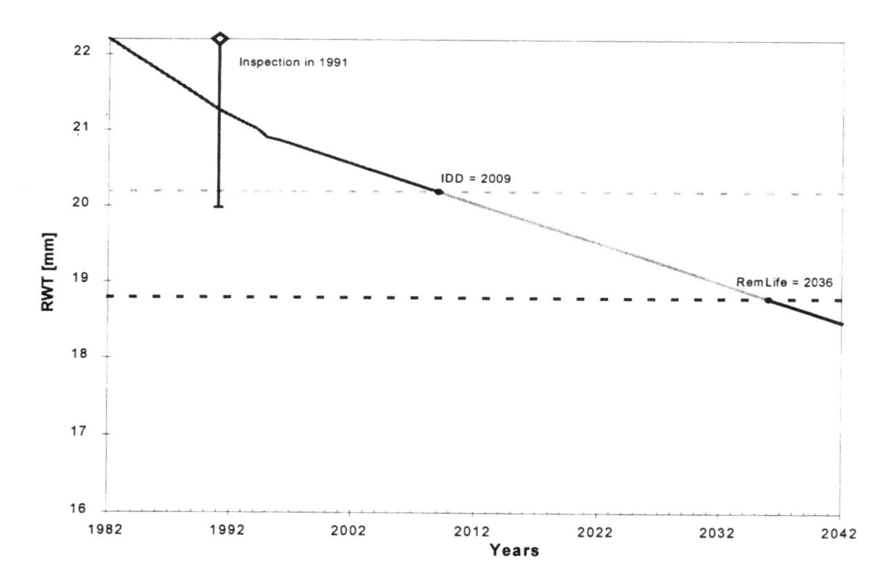

Figure 4: Remnant Life analysis for offshore 36-inch gas pipeline. (RWT = Remaining Wall Thickness, IDD = Inspection Due Date)

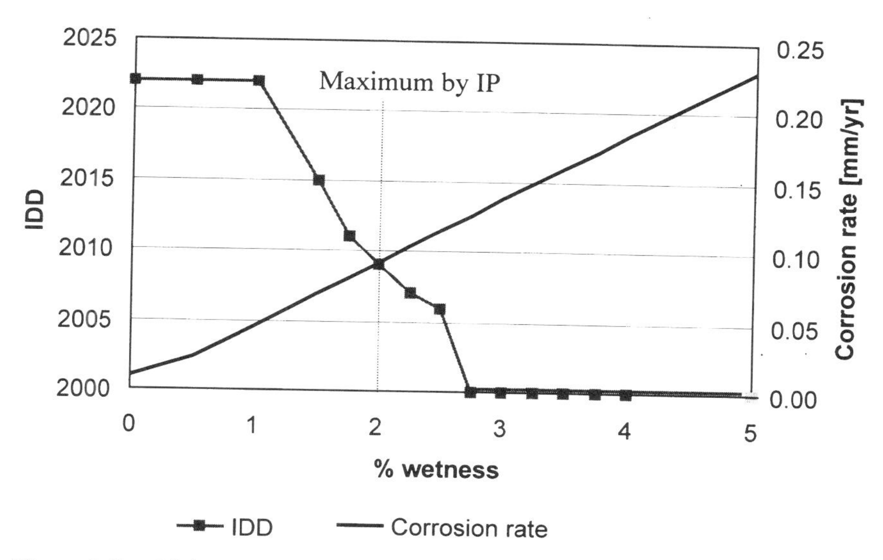

Figure 5: Sensitivity analysis on % wetness for offshore 18-inch gas/condensate pipeline. (Note that the maximum remnant life is cut off at 50 years resulting in a maximum inspection interval of 25 years for this pipeline)

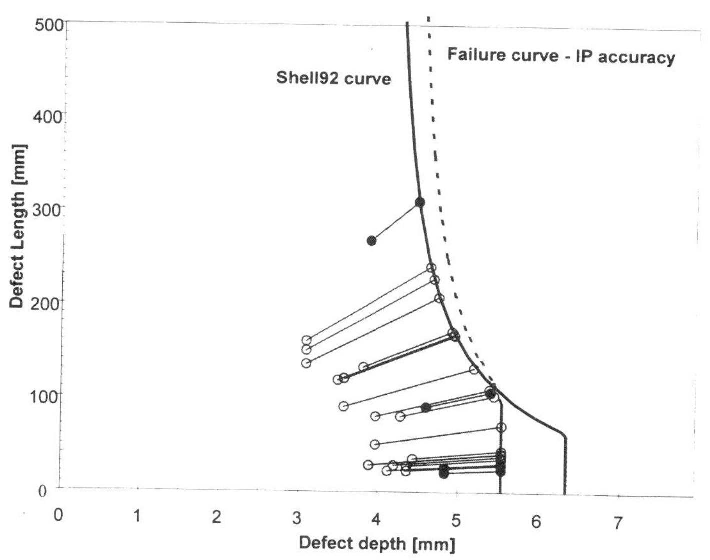

Figure 6: ASSESS-PIPE analysis for onshore 20-inch oil pipeline by which the corrosion tolerance of each defect is calculated.

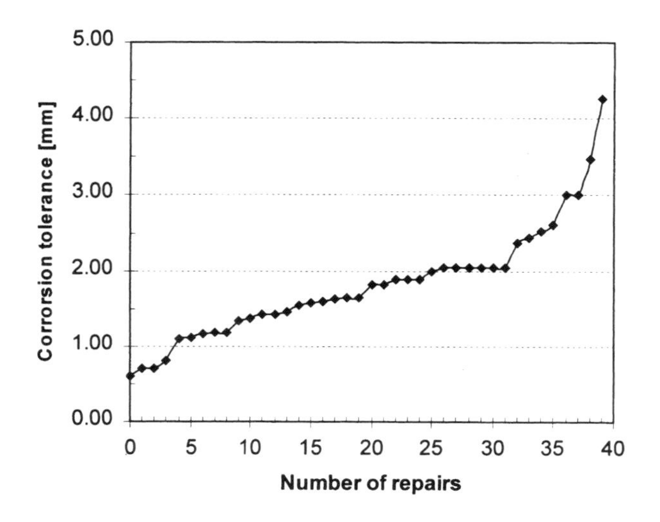

Figure 7: Corrosion tolerance as function of number of repairs for onshore 20-inch oil pipeline.

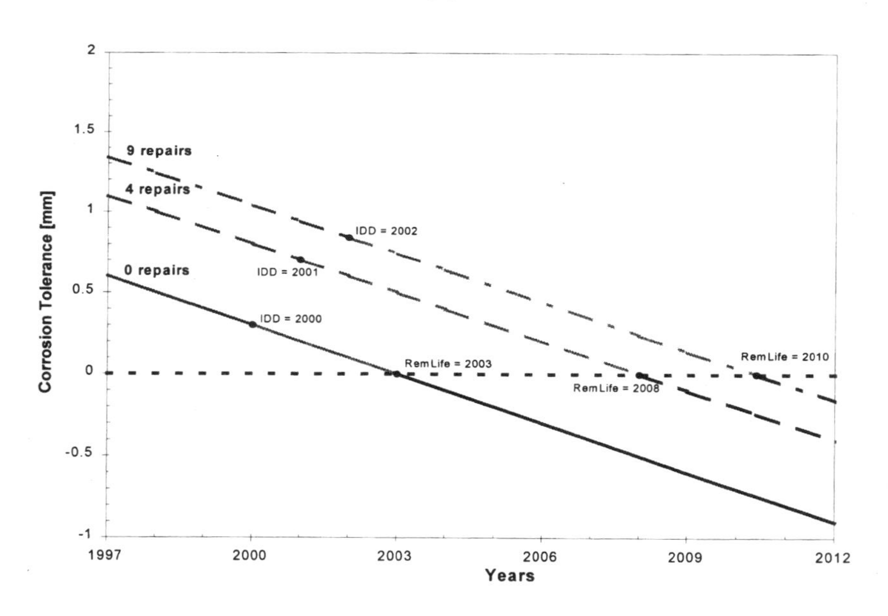

Figure 8: Remnant Life analysis for onshore 20-inch oil pipeline in case of respectively 0, 4 and 9 repairs.

C571/030/99

The use of geographical imaging technology for improving onshore pipeline asset management activities

A FRASER
Integrated Statistical Solutions Limited, Sunderland, UK
R W OWEN and **K SAHIN**
BG Technology, Loughborough, UK

ABSTRACT
Increasingly common issues facing the operation of ageing onshore pipelines include;

- Pressure on the pipeline's encroachment corridor due to increasing urbanisation. Creating a requirement for increasingly stringent and auditable dwelling surveys.
- Increased threat from third party damage through the intensification of farming and construction activities.
- Increased traffic on rail and road networks.
- Increased demand for pipeline uprating to provide extra capacity for gas transportation throughput and diurnal storage.

A common factor in managing these issues is the requirement to collect and integrate large amounts of detailed data concerning a pipeline's construction, and operating conditions. In addition there is a requirements to gather information concerning the pipeline's ground coverage, location of roads and rail networks, houses and industrial units, and where changes have occurred in these since the time of construction.

Compiling and integrating this data has proved to be highly complicated and time consuming due to the disparate way in which data has been compiled over many years of a pipeline's operational life. Given the age of many onshore pipelines, much of this relevant data has been stored in a non-digital form. These problems have led BG Technology and Integrated Statistical Solutions Ltd to develop a data management system for capturing, and integrating data, and allowing the operations outlined above to be handled efficiently. The system is designed around the use of Geographical Imaging, Geographic Information Systems (GIS) and relational database technology, using industry standard software tools Erdas Imagine, and ArcInfo. The system allows air photographs, and the new Very High Resolution (VHR) satellite imagery to be integrated with pipeline construction and operations data, and furthermore allows dwelling encroachment, infringement, pipeline construction changes and pressure up-rating studies to be carried out within a single software environment.

This paper describes the system and gives examples of its use. In addition, the paper outlines potential improvements that are planned to the system in the future.

1. INTRODUCTION

The operational management of onshore pipelines is a complex process, involving the combination of activities designed to ensure both the safety and operational efficiency. This is particularly the case for high pressure transmission pipelines. Typically, onshore pipeline asset management involves access to, and processing of many diverse forms of data, typically ranging from pipeline inspection data through to information concerning population density surrounding the pipeline. Historically, because of this diversity, such data has been recorded and held in many different ways, including paper records, microfiche, and widely disparate digital databases running on, in many cases, non-compatible computer systems. This situation has led to many pipeline asset management tasks being inefficient This situation is exemplified in the areas of pipeline surveillance, population estimation, and pipeline routing. All of these tasks rely heavily on the use of imagery. The problems associated with the operational handling of large quantities of geographic image data have until recently impaired the use of this technology, and has typically resulted in the use of video and paper air photographs. Both of these have the drawback of not being able to make quantitative measurements of the distance between the pipeline, houses and roads.

This situation has been thoroughly reviewed by Integrated Statistical Solutions and BG Technology, [1]. The results of this review have led Integrated Statistical Solutions Ltd. (ISS), and BG Technology to develop a data management system for capturing and integrating geographical imaging data and pipeline information, and for allowing a number of pipeline asset management operations to be handled with greater efficiency and accuracy.

This paper describes in detail the system that has been developed. In particular, the functionality of the system, and its underlying architecture and design strategy are described. The paper describes in detail some of the operations that a pipeline operator, and in this case Transco typically have to perform. This description also includes indications of where current practice is regarded as inefficient, or where there is scope for improvement. Finally, an analysis is made of the current and future effectiveness of this approach, taking into consideration advances in computer software and hardware processing performance, and in the variety and quality of geographical imaging data, and particularly remote sensing data.

2. ISSUES FACING OPERATORS OF AGEING ONSHORE PIPELINES.

There are a number of issues facing operators of ageing onshore pipelines that have been relevant to the development of the pipeline asset management system, and geographical imaging technology reviewed in this paper. These are briefly described below;

2.1 Demographic change monitoring

The majority of the UK's high pressure gas transmission network was initially routed through rural areas containing low urban population densities. Many cities have since undergone significant growth, with the emphasis on 'out of town' developments on their outskirts. In many cases this has resulted in increased population densities in the vicinity of pipelines. This has been a particular problem in the South East of the UK. This phenomena necessitates increased population density monitoring by operators. The update of traditional mapping information, such as that supplied by the Ordnance Survey in the UK does not, in many instances keep pace with this urban development. There is therefore an acknowledged reliance by operators on the use of remote imagery to quantify and update population densities in the vicinity of pipelines.

2.2 Increased road use.

The use of cars and commercial road haulage has increased dramatically in recent years. As a result, 'as built' estimates of road traffic density at pipeline crossings, and on roads close to pipelines can be significantly lower than today's values.

3. TYPICAL PIPELINE OPERATIONS INVOLVING GEOGRAPHICAL IMAGING

This section describes in detail one particular operational task that has to be carried out by onshore pipeline operators. (In the example by Transco.) This task relies heavily on the use of remote imagery to estimate population density. It is this task that has formed the primary application for the system described in this paper.

3.1 Verification of Maximum Permissible Operating Pressure

Transco operates nearly 19,000 km of high pressure (> 7 bar) gas transmission pipelines that transport gas from the import terminals to the lower pressure network system. Transco transmission pipelines are generally designed, constructed and operated according to the recommendations in the Institution of Gas Engineers document IGE/TD/1 [2]. This recommends that an audit of the pipeline should be carried out at intervals of not more than four years to confirm the Maximum Permissible Operating Pressure (MPOP). The MPOP is the maximum pressure at which a pipeline may be operated, either by design or by down-rating or up-rating in accordance with the IGE/TD/1 recommendations, as a result of a change from the original design conditions, or because of temporary constraint.

One of the issues that can, with the passage of time, limit the MPOP is infrastructure development, adjacent to the pipeline, which alters the proximity and/or population density. Measurement of population density may be based on a survey, for example by aerial photography, of normally occupied buildings and premises where people congregate for significant periods of time, for example schools, public halls etc. The population density surrounding the pipeline determines the area classification, which will require different design criteria, with particular reference to operating stress level and proximity. Area types are classed as Type R, which are rural areas with a population density not exceeding 2.5 persons per hectare, and Type S, which are suburban areas, with a population density exceeding 2.5 persons per hectare. Particular attention also needs to be given to pipelines crossing, running along or under traffic routes. There are two categories of traffic route, depending on traffic density, which will again require different design criteria.

Proximity and population density infringements may be found during routine pipeline surveillance, but compliance should be verified as part of the four yearly audit. Traditionally, the first examination has been undertaken using up to date geographical population data in the form of black and white aerial photographs, together with paper strip maps (1:2500) of the pipeline. Subsequent inspection has usually been carried out using a mixture of video and land vantage point surveys. These procedures have tended to be labour intensive and have often required site visits to verify information and to accurately measure distances from the pipeline.

4. TECHNICAL OPPORTUNITIES

There are a number of technical developments that have occurred in recent years that can improve the handling of information, and particularly geographic imaging data, [3], [4]. These technical developments are;

4.1 Remote imagery

Digital imagery, both from aircraft and satellites is becoming more widespread, cheaper, and of better quality. Furthermore, satellite data will soon be available on a global basis at a resolution capable of identifying individual dwellings, and cars. An example of this is shown in figure 1. This shows imagery from the Russian KVR satellite. The resolution of this imagery is 2m. New satellite systems will deliver data at considerably higher quality and at a greater resolution. It should also be pointed out that there will be a number of additional features to this data, namely;

(a) Collection of Very Near Infrared (VNIR) imagery capable of very sensitive detection of leakage, crop loss estimates and improved building identification.
(b) Ability to collect data every 1-3 days anywhere in the world.
(c) The ability to acquire data quickly, without recourse to lengthy planning and Quality Control procedures.

4.2 Geographic imaging software

There has been a significant improvement in the functionality of geographic imaging software. As a result, it is now practical for non-imaging experts to acquire and process remote imaging data, and integrate it with existing pipeline asset data.

4.3 Database technology

Common standards and formats now exist for the transfer of data between different databases running on different machines and operating system environments. As such, the rapid access and integration of data is now practical and relatively simple.

4.4 Data processing and mass storage technology

Until recently, systems involving the processing and integration of digital imagery, mapping and pipeline asset information would have required the use of highly complex and expensive workstation technology. This level of operation is now possible on an advanced PC, or even a Laptop. In addition, the ability to store large quantities for image data is now relatively inexpensive due to the introduction of low cost high capacity disk systems.

Figure 1. Satellite imagery taken from the Russian KVR ex-spy satellite

These developments offer a significant opportunity to operators in that, combined, they offer the potential of developing very cheap, highly integrated pipeline asset management software tools requiring a low level of Information Technology maintenance. It is precisely this potential that the system described in this paper is addressed at.

5. THE ISS – BG TECHNOLOGY APPROACH

The system described here has been developed to assist in the four yearly audit of the pipeline to confirm the MPOP, in particular, assessing the infrastructure development, adjacent to the pipeline, which can change the proximity and/or population density. The system currently enables the engineer to use digital geo-referenced strip maps, which can be overlaid with digital colour aerial imagery. All pipeline details are also readily available through the accompanying pipeline database, which includes relevant pipeline attributes, such as diameter, wall thickness, pipe type, material grade, protection etc. The system can then establish area type boundaries, and identify any proximity and/or population density infringements. The benefits of such a system will be greater at the time of the next audit, when all pipeline information is already available in the system, and any changes identified by the aerial photography can be easily assessed. However, of greater significance, is the extension of this type of system to a broader pipeline asset management system. A variety of different types of information, which are important to the management of the pipeline, can be easily displayed in relation to its geographical position on the pipeline. For example, risk assessments can be stored and referenced by location to support the four yearly pipeline audit, together with geo-technical information on ground conditions, locations of pipeline repairs or location of stress analyses previously undertaken.

5.1 System Design

Figure 2 shows the process by which raw data, either recently captured or residing in an

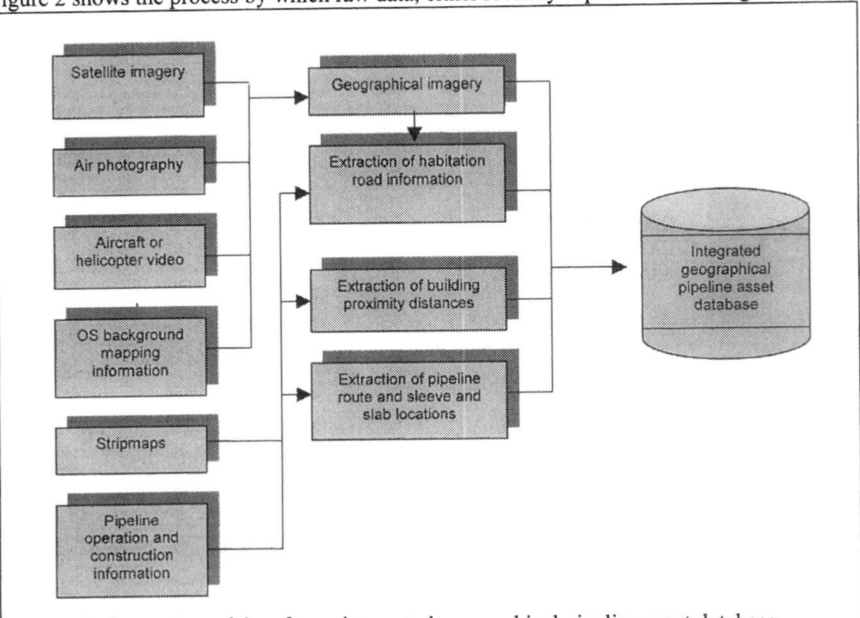

Figure 2. Generation of data for an integrated geographical pipeline asset database.

existing database, is imported into the system, and converted into information that is suitable for pipeline operations. An important aspect of this is the creation of a database containing all of the pipeline information necessary for confirming the MPOP. The system can import imagery from a wide range of sources including analogue air photography, digital air photography, video frames, and a wide variety of satellite imagery. The system contains a comprehensive range of image processing tools, that converts the imagery into an accurate mapping source. The 'Geographic Imagery' can now be readily combined with other geographical data such as strip maps, to show accurate information concerning buildings, roads etc, and allow calculations of distances between dwellings and a pipeline to be made.

The system has the inherent functionality to import both paper and digital stripmaps, and again process them in such a way as to allow integration with geographical imagery and other pipeline information. The crucial aspect of stripmap integration is the system's ability to extract pipeline location, and sleeve positions from the stripmap, and place these in the system's database.

The system provides tools to extract road crossings and building positions from the geographical imagery, and allows these to be input into the database. This allows automatic calculation of building to pipeline distances and hence population densities.

Figure 3. shows how the integrated database is used by the system at present, and how in future, the same database can be used to carry out additional pipeline asset management tasks To date, the IGE/TD/1 design criteria, in terms of population density, design factors and

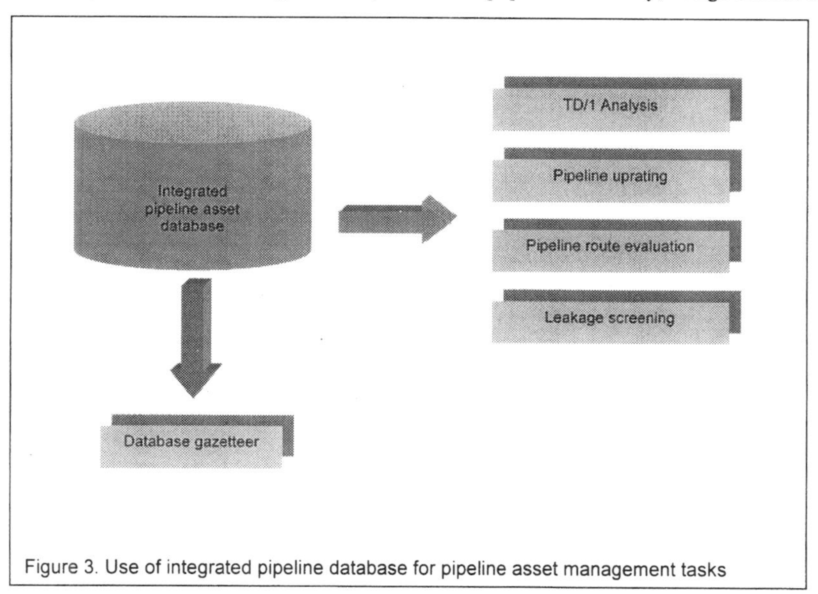

Figure 3. Use of integrated pipeline database for pipeline asset management tasks

road/rail crossings have been incorporated into the system. This allows the user to access data to carry out;

- Automatic generation of 1 and 4 Building Proximity Distance (BPDs) corridors around the pipeline.
- Generate population densities along the pipeline.
- Analyse road and rail crossings, and examine pipeline wall thickness and protection sleeving in these areas.
- Automatically assess whether the IGE/TD/1 design criteria (in terms of population density, design factors and road/rail crossings) have been met, and if not, identify locations where the infringements have occurred. An important aspect of this is the user's

ability to assess what remedial measures are required in order to meet IGE/TD/1 design criteria . In effect, the system allows the user to carry out pipeline redesign within the system.

Figure 4 shows a typical example of how the system operates. This shows an air photo overlaid with a stripmap. In addition, it shows how the system calculates the 4 BPD corridor based on the pipeline operation parameters. Finally figure 4 shows how the system holds and displays the locations of habitation which are used to assess urban densities along the pipeline route.

In order to report results, the system provides a comprehensive reporting interface. The output report contains details of all BPD, population density and road and rail crossing violations.

5.2 System Architecture

The approach taken to the design of this system has been to maximise the use of standard software tools. Because of this, the system has been designed and built quickly, whilst maintaining a high level of flexibility. As a result system enhancements can be made quickly, and code maintenance can be kept to a minimum. The system is designed around the use of geographical imaging, GIS and relational database technology, using leading industry standard software tools;

- Erdas Imagine (Geographic Imaging)
- Arc/Info (GIS)
- Dbase (Database)

These packages are inherently highly integrated, allowing geographical imagery, pipeline engineering information, and geographical data, such as OS mapping, pipeline location, roads, and habitation to be readily accessed and used together by a user with minimum effort. A great deal of attention has been applied to the design and implementation of the pipeline asset database. Detailed data models and relationships have been constructed to hold engineering data associated with individual pipeline sections, such as pipe diameter, wall thickness, pipe material, design factor, MPOP etc. For the population, the database holds information on all dwellings surrounding the pipeline, and estimated numbers of inhabitants. The holding of geographic imaging in the database is significant in giving the user the ability to carry out rapid change control, and reuse imagery for other operations such as installation siting, pipeline routing and pressure uprating and leakage surveys. In summary, correct database design has been key to the design of a system with the functionality to assist in carrying out IGE/TD/1 surveys and having the flexibility to be extended to carry out a number of other pipeline asset management operations reliant on the integration of geographic imaging and ancilliary pipeline and mapping information.

6. RESULTS OF SYSTEM FIELD TRIALS

The field trial was carried out in the Transco NTS offices. The three pipelines chosen were in the Bishop Auckland area, with a total length of about 100 km. Pipeline strip maps for the three pipeline sections were imported from the Transco Digital Records System (this is a graphical pipeline database) and colour aerial photography of the trial area was commissioned (to be supplied in a digital form). The system allows the display of all of the available data for the area of interest, such as all of the strip maps and photographs that are in the system. After initial training, Transco pipeline engineers used the system to assist in carrying out the IGE/TD/1 four yearly survey. The system was found to be generally user friendly, as most of the functionality was implemented graphically. Some enhancements were identified by the Transco users, which are currently being implemented. However, the system, which was developed initially as an aid to the IGE/TD/1 survey, was found to be a useful tool, although would be even more useful as part of a broader pipeline asset management system.

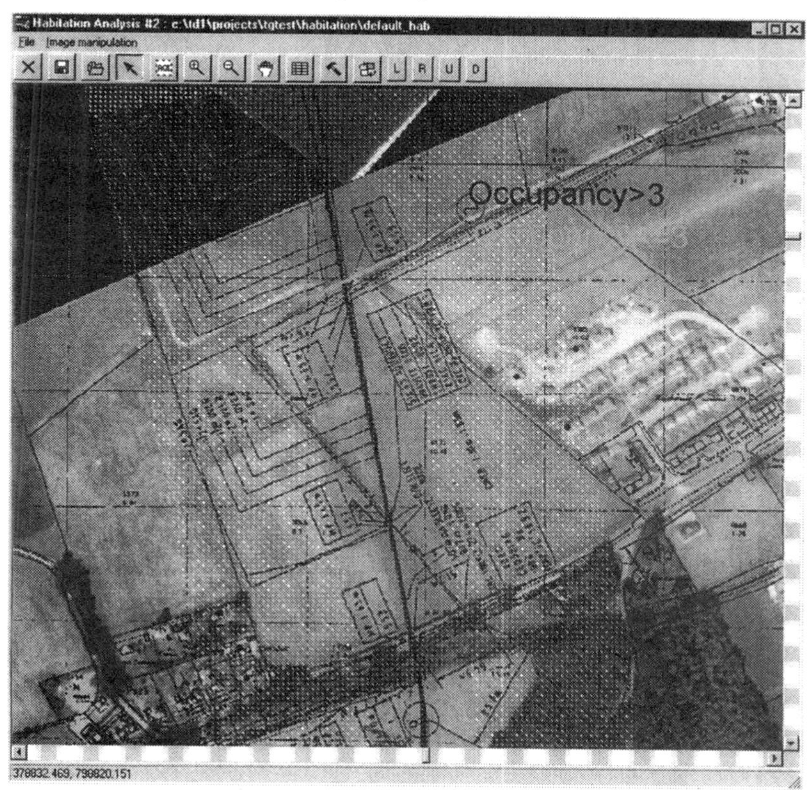

Figure 4. Example of integration of geographical imagery and stripmap data

There were a number of observations that were drawn concerning the use of the system in a future operational context. Firstly, there is an initial overhead concerning the compilation of data to put into the system. This overhead can typically add 20-30% to the cost of survey. However, this overhead occurs only at the first use of the system. Subsequent survey times and cost are significantly reduced due to the digital nature of the data. In addition, the pipeline asset database compiled as a result of this exercise can readily bring cost savings to other pipeline integrity operations. The overhead associated with data input will reduce radically in the near future as base level information, such as stripmaps and pipeline location become increasingly available in digital form. There is a point to be made concerning existing stripmap information. This data has been collected from ground surveys and a series of manual drafting. This data is open to error. In the near future highly accurate thermal imagery will become available that will allow the pipeline's position to detected, providing a quick and accurate confirmation of as built pipeline location records. Finally, the overhead associated with data entry, does currently create a bottleneck in the use of the system operationally. This bottleneck will be eliminated in the near future due to the factors described above, and the emerging availability of digital imagery that is geographically tagged automatically using GPS technology. This will remove all of the geocoding time currently associated with preparing the data for analysis.

7. CONCLUSIONS

There are a number of conclusions that can be drawn concerning the use of geographical imaging technology for improving onshore pipeline asset management operations. Firstly, geographical imaging technology is now a cheap technology, which is operational on standard PC environments, and even field laptop systems. Geographical imaging data is becoming much more cost effective, a process that is set to accelerate with the imminent introduction of very high resolution satellite imagery. Furthermore, aircraft and satellite imagery is increasing in quality, thereby enhancing its capabilities for use in many pipeline operations. This concept of 're-use' or 'multiple-use' if used correctly, significantly improves the cost/benefit associated with the integrated approach presented in this paper. This is also true of re-evaluation of a pipeline after the 4 yearly affirmation of the MPOP. The highly integrated form of the data collected for the initial study results in rapid re-surveying in subsequent years, and hence a long term cost saving.

Secondly, Calculations carried out using the system presented here have been thoroughly tested, and have been proven to be highly accurate. The calculations are also consistent, independent of operator. Using the system, operators can carry out surveys of large pipeline lengths; up to 80kms in a single project.

Finally, the system is capable of being used to assist in the design of new pipelines, and for rapid evaluation of the effects of changes in operational parameters, such as pressure uprating.

8. REFERENCES

[1] Fraser, A.J., 'Digital Image Processing as an aid to IGE/TD/1 Four-yearly surveys. Review of current availability of images and demographic databases'. Report compiled for BG Technology by Integrated Statistical Solutions Ltd. April 1997.

[2] 'Recommendations for transportation and distribution practice' IGE/TD/1. Institute of Gas Engineers.

[3] Fraser, A.J., 'Use of high resolution satellites for the integrity management of onshore transmission pipelines.' Report produced by Integrated Statistical Solutions Ltd. for Andrew Palmer Associates Ltd. 12th February 1998.

[4] Fraser, A.J., 'Final Report: A Study into the use of Digital Airborne Imagery for Improved Encroachment Analysis and Hazard Assessment.' Internal report prepared for BG Transco by Integrated Statistical Solutions Ltd. 20th January 1997

C571/031/99

Integrity assessment and inspection planning of corroded pipelines using DNV RP-F101

O H BJORNOY, G SIGURDSSON, T SYDBERGER, and **M MARLEY**
Det Norske Veritas, Hovik, Norway

ABSTRACT

Integrity assessment and inspection planning of corroded pipelines using the **DNV Recommended Practice, "RP-F101 Corroded Pipelines"** (1) are described in this paper. The basis for the DNV RP-F101 was developed in co-operation between BG Technology and Det Norske Veritas (DNV). The development has been sponsored by several oil companies and regulatory agencies through sponsorship of joint industry projects.

RP-F101 gives acceptance criteria for a uniform reliability level of corroded pipelines with different pipe materials, pipe geometries, and defect sizes. Different inspection methods with various sizing accuracy are treated. The equations for determination of the allowable operating pressure are calibrated using probabilistic methods to ensure a uniform reliability level, in harmony with the design philosophy in the DNV Pipeline Standard (2,3).

The methods in RP-F101 were developed for integrity assessment of pipeline with corrosion defects, and are valid at the time of inspection. However, by its nature, a corrosion defect will normally develop with time, and extension of the methods in RP-F101 to serve as a basis for decision making, e.g. for inspection planning, is described.

1 BACKGROUND

When severe corrosion is observed in a pipeline, the decision of actions to take should be based on an assessment that accounts for uncertainties, both in the inspection measurements and in the capacity evaluation. If the integrity of the pipeline can not be demonstrated based on the available inspection results, the decision of further actions should also take into account the uncertainty of possible further inspection results. Costly repair, replacements or other actions should be avoided if more accurate inspection results could demonstrate acceptable integrity.

The equations in RP-F101 account directly for the accuracy in sizing the corrosion defect. Hence, the possible benefit –increased allowable pressure – obtained by improving the accuracy of the inspections can be seen immediately. Or, for a given pressure, larger measured defect size is allowed by improving the inspection accuracy.

For submarine pipelines, the actual dimensions of a significant internal corrosion damage detected by pigging cannot readily be determined by external NDT as for pipelines on land. Furthermore, contrary to the external corrosion of the latter, the control of future development of internal corrosion damage is not a straightforward task and accordingly its efficiency must normally be verified by a new inspection. In order to assess the integrity of a pipeline with internal corrosion, it is therefore necessary to estimate the growth rate of the existing corrosion damage.

2 RP-F101 CORRODED PIPELINES

The DNV RP-F101 Corroded Pipelines provides recommended practice for assessing pipelines containing corrosion defects. The recommendations include single defects, interacting defects and complex shaped defects for a consistent reliability level for various pipeline dimensions, loading and defect size. The target reliability levels are dependent on the consequence of a potential failure, and are determined by the content and location of the pipeline and characterised as Safety Class. There are three Safety Classes in the DNV Pipeline Standard; Low, Normal and High, with associated target reliability levels. Oil and gas pipelines are normally classified as Safety Class Normal, and Safety Class High for risers and parts of the pipeline in areas where frequent human activity is anticipated. Safety Class Low can be considered for water pipelines.

As the equations account for the uncertainties in the sizing of the defect, the inspection accuracy is an important input parameter, and should preferably be selected in consultation with the inspection tool supplier. The deterministic approach given in RP-F101 includes equations with partial safety factors that depend on the Safety Class classification and the inspection accuracy. This ensures a safer assessment when the consequence of a failure is high, and the inspection accuracy is taken into consideration in the determination of the partial safety factors. During the development of RP-F101, the partial safety factors in the acceptance equations were determined by applying probabilistic code calibration procedures using full distribution reliability methods (4,5). The users have only to select the Safety Class and the sizing accuracy, and then extract the partial safety factors from tables in RP-F101. The acceptance equation, including the extracted partial factors, give a consistent reliability level for all defect sizes. The users do not have to explicitly account for the sizing uncertainty for each single defect.

Acceptance equation

The allowable pressure for a pipeline with a single corrosion defect is given by

$$p_{corr} = \gamma_m \frac{2\,t\,SMTS}{(D-t)} \frac{\left(1-\gamma_d (d/t)^*\right)}{\left(1-\frac{\gamma_d (d/t)^*}{Q}\right)} \leq p_{mao}$$

where

$$Q = \sqrt{1 + 0.31\left(\frac{l}{\sqrt{Dt}}\right)^2}$$

$$(d/t)^* = (d/t)_{meas} + \varepsilon_d \mathrm{StD}[d/t]$$

and

D	=	outer diameter
d	=	depth of corrosion defect
t	=	wall thickness (nominal)
l	=	measured length of corrosion defect
$(d/t)_{meas}$	=	measured relative corrosion depth
p_{mao}	=	maximum allowable operating pressure
γ_m	=	partial safety factor for model prediction and safety class
γ_d	=	partial safety factor for corrosion depth
ε_d	=	factor for defining a fractile value for the corrosion depth
$StD[d/t]$	=	standard deviation of the measured (d/t) ratio (based on the specification of the inspection tool).
$SMTS$	=	Specified Minimum Tensile Strength

Partial safety factors and fractile values are derived from DNV RP-F101. The allowable operating pressure of a pipeline with a corrosion defect measured to the same size is highly dependent on the sizing capabilities of the inspection tool. This effect is illustrated in Figure 1, which shows allowable pressure for four levels of inspection accuracy: exact, and ± 5, 10, and 20% of wall thickness (in each case with 80% confidence). A measured defect depth of 50% of wall thickness was selected, and is in the figure given as a function of the defect length. All lines are for Safety Class Normal, which correspond to a reliability level of 10^{-4}. For the inspection accuracy of ± 20% this reliability level can not be demonstrated for any pressure, and hence the allowable pressure is zero. The importance of the sizing capability of the inspecting tool is clearly illustrated.

The major challenge in the method outlined above is to quantify the uncertainties in the defect sizing. The uncertainties associated with the inspection should be based on the specification of the sizing capabilities supplied by the tool operator. It is not straightforward for the operators to quantify the sizing accuracy, as the accuracy is dependent on many factors, but this parameter is obviously very important. One way the accuracy has been expressed in inspection reports is for instance "±10% of the wall thickness with 80% confidence". For ease of use, in RP-F101 the standard deviations are tabulated for selected values of accuracy and confidence level based on this kind of expression. In these cases a normal distribution has been assumed.

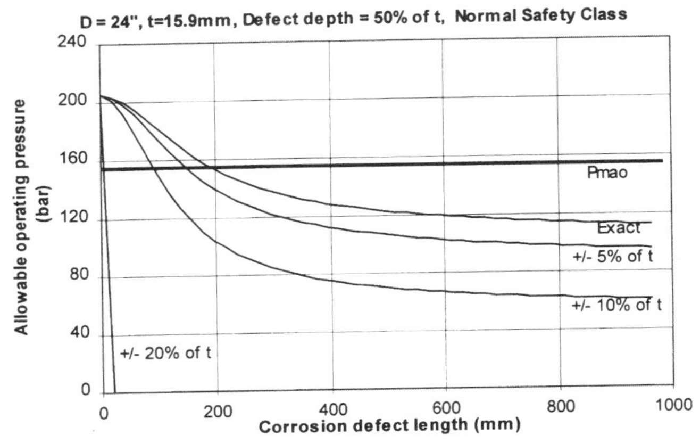

Figure 1 Allowable operating pressure vs. inspection

The equations in RP-F101 may also be presented in terms of an allowable measured corrosion depth and length for a given, and fixed, operating pressure. The allowable measured corrosion depth is highly dependent on the inspection accuracy. In Figure 2, the allowable measured defect size for safety class Normal is plotted for various inspection sizing capabilities, for an operating pressure of 155 bar. As long as the measured defect size is below the appropriate line, the failure probability of the defect is less then the target failure probability, and is therefore acceptable. The criterion given in the figure below is valid at the time when the inspection was performed. If the corrosion is not stopped and the defects develop, the time until the defects become critical should be evaluated. This is considered in the next section.

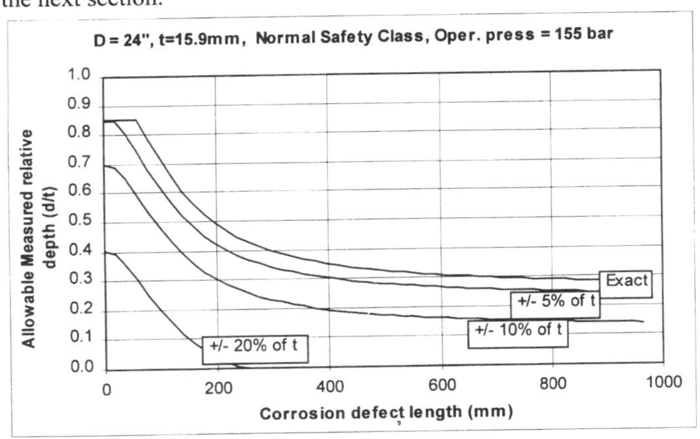

Figure 2 Allowable measured defect depth/length for given operating pressure and inspection accuracy.

3 INSPECTION PLANNING AND REMAINING LIFE TIME ASSESSMENT

The assessment described on the previous chapter is valid at the time of the last inspection, and there is no inherent additional safety to account for further development of the corrosion defects. However, the methodology may be extended, and together with an estimate of the future corrosion rate, both depth and longitudinally, form the basis for evaluation of the integrity of the pipeline in the future. This evaluation can be used for planning for the next inspection or the assessment of the remaining lifetime.

In the following, a methodology for inspection planning of a corroded pipeline is described for the case where the criterion is given as a failure probability in accordance with the safety philosophy in DNV Pipeline Standard, reflected in the safety class characterisation. This criterion, together with the DNV RP-F101 model, inspection data and a qualified estimate of the future corrosion development, form the basis for the inspection planning or the remaining lifetime of the corroded pipeline

In the previous section the importance of the sizing uncertainties of the corrosion defect for the allowable operating pressure was demonstrated. When evaluating the future integrity of a corroded pipeline, the estimated corrosion rate with its associated uncertainties also add to the sizing uncertainties and should be considered. Detailed probabilistic methods can be adopted, but in many cases it is sufficient to base the assessment on the approach in RP-F101, in which the uncertainties in the defect sizing is accounted for in the partial safety factors.

For pipelines affected by internal corrosion, the further development of the most severe damage should be predicted based on an assessment of future corrosion rates both in the thickness direction (i.e. reducing the wall thickness) and parallel to the pipe axis (i.e. longitudinal growth). If only one inspection has been performed, the corrosion rate may be assumed to have been constant from the start of operation to the time of the inspection. Improved estimates may be based on results from two consecutive pipeline inspections. For either case, it is obvious that such estimates are affected by the sizing capability of the inspection tool used.

However, it is highly recommended that the initial assumption of a constant corrosion rate is subsequently assessed by a corrosion specialist. Hence, knowledge of the actual corrosion mechanism and how it is affected by past and future operational parameters (e.g. water cut, temperature, chemical treatments, etc) may justify adjustments of corrosion rate estimates as described above. This may lead to the initial estimate being either enhanced or reduced. It should be obvious, however, that the assessment of future corrosion will be subject to considerable uncertainty, the magnitude of which has to be assessed in order to evaluate the reliability of a corroded (and still corroding) pipeline.

For pipelines inspected by for instance MFL (Magnetic Flux Leakage) intelligent pigs, the corrosion defects are defined by maximum corrosion depth (relative to the wall thickness) and corrosion defect length. The results can be presented in a figure with axis of measured corrosion depth and corrosion length, and with a full line indicating the allowable measured defect size, see Figure 3. For illustration, examples of corrosion defects are included in the figure. The sizing accuracy is ±10% with 80% confidence level. The dotted line indicates the allowable measured defect size if the defect sizes were exactly known.

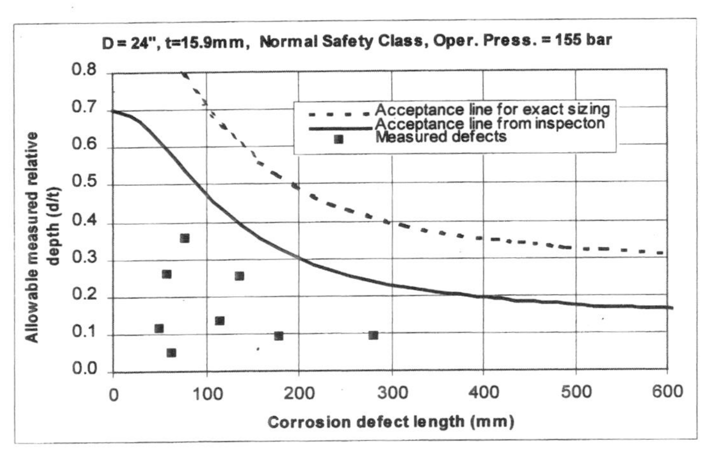

Figure 3 Allowable defect size (with given operating pressure and inspection accuracy)

In inspection planning or estimation of the remaining lifetime the development of the corrosion defects needs to be assessed. The corrosion depth d_T and length l_T at time T may be expressed as, see also (6).

$$d_T = d_0 + T \cdot r_{corr}$$

$$l_T = l_0 + T \cdot r_{corr,length}$$

where

d_T = corrosion depth at time T

d_0 = measured corrosion depth

T = time interval in years from last inspection

r_{corr} = average annual corrosion rate

l_T = longitudinal defect length at time T

l_0 = measured longitudinal defect length

$r_{corr,length}$ = average annual corrosion rate in the length direction

To estimate the uncertainty in the corrosion defect size at time T, the uncertainty in the future corrosion development must be added to the uncertainty of the defect size from the inspection. Hence the uncertainty increases with time. Since the uncertainty in the corrosion depth at time T is much higher that the uncertainty in the pipe wall thickness, the expected value and the standard deviation of the relative corrosion depth may be approximated as:

$$\left(\frac{\bar{d}}{t}\right)_T \approx (d/t)_0 + T \cdot (\bar{r}_{corr}/t)$$

$$StD[d/t]_T \approx \sqrt{StD[(d/t)_0]^2 + \frac{T^2}{t^2} \cdot StD[r_{corr}]^2}$$

where

$(d/t)_0$	=	measured relative corrosion depth
$\bar{r}_{corr}$	=	expected average annual corrosion rate
$StD[r_{corr}]$	=	standard deviation of the average annual corrosion rate

Dependent on the corrosion mechanism, the corrosion defect can develop differently in depth and length, and should preferably be estimated by a material and corrosion specialist in each case. A simple and easy alternative is to assume that the length grows in proportion with the depth. Then the length at time T can be expressed as:

$$l_T = l_0 \cdot \left(1 + \frac{T \cdot r_{corr}}{d_0}\right)$$

The capacity at the future time, T, is calculated from the equations in RP-F101 with

$(\bar{d}_r / t)$ is used for $(d/t)_{meas}$,
l_T is used for l, and
$StD[d/t]_T$ is used for $StD[d/t]$.

The future corrosion rate with associated uncertainties expressed as standard deviations has to be estimated and should preferably be performed by a material and corrosion specialist. The estimate is based on the measured corrosion, and expected future operational conditions. The corrosion specialist should provide a best guess rate and an indication of maximum (and minimum) corrosion, together with an estimate of how certain these are. With this information, the standard deviations in the rate can be estimated.

Figure 4 extends the example shown in Figure 3 from time T_0 to time T_1. The defects have increased in length and depth. The capacity (ie, the allowable defect size) at time T_0 is indicated by the uppermost curve.

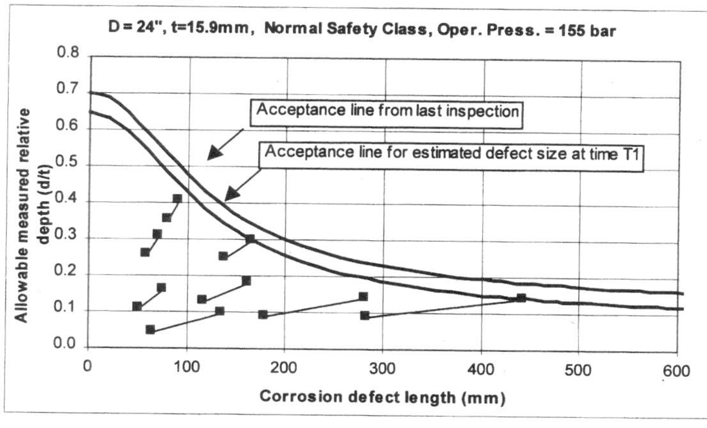

Figure 4 Allowable defect size at future date, T_1.

This line depends on the uncertainties in the defect sizes, which at time T_0, immediately following an inspection, depends only on the inspection accuracy. At time T_1, the standard deviation of the size of the defect has increased, due to the additional uncertainty in the estimate of the future corrosion rate. Hence the allowable defect size is reduced at time T_1. The partial safety factors as a function of defect size uncertainty are given in RP-F101, and these are used to obtain the updated allowable defect size curve. By updating the acceptance lines and plotting the estimated increased defect size, it can be seen directly from the plot when the acceptance criterion is exceeded. The increase of the corrosion defect uses the expected value of the average annual corrosion rate, while the additional uncertainties associated with the defect size is accounted for in the acceptance line, which makes this method easy to apply. In (7) an example with detailed calculations is shown.

When a defect exceeds the acceptance line, the criterion for allowing further operation of the pipeline is exceeded, and an action is required to allow further operation. There are several options for further operation:

- inspection in order to reduce the uncertainties in the corrosion rate estimate
- repair or replacement of the critical sections;
- reduced pressure or other operational constraints;
- more advanced assessment of the defects; or
- more accurate inspections in order to reduce the overall defect size uncertainty.

Note that inspecting with greater accuracy reduces sizing uncertainties and will increase the acceptable measured defect size for a prescribed operating pressure and safety level.

Example

If the selected action is to perform an inspection in order to confirm the integrity of the pipeline the most severe defects will determine when a new inspection should be performed. In addition to the representation in Figure 4 where all defects are shown, the most severe defects can be represented as allowable pressure versus time as shown in Figure 5.

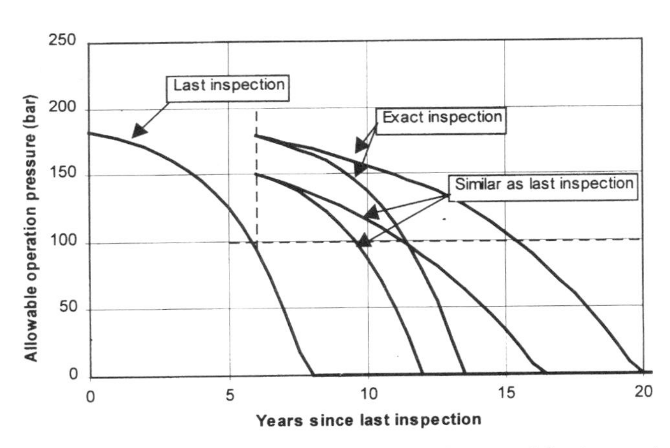

Figure 5 Allowable pressure versus time used for inspection planning

The example given in Figure 5 is for a 24" (609mm) pipeline with 15.9 mm wall thickness of X60 grade material, and operated at 155 bar. An inspection was carried out with using a MFL (Magnetic Flux Leakage) intelligent pig with a quoted accuracy of ±10% of the wall thickness with 80% confidence. The most severe corrosion defect was measured to 30% of the wall thickness and 100mm long. Safety Class Normal is selected, and the additional material requirements are not specified (2,3). According to RP-F101 the allowable operating pressure is 182 bar at the time of inspection (0 at the x-axis).

It is however expected that the defect will develop with time both in depth and length. The corrosion rate is estimated to 0.3 mm/year in depth, and the uncertainty in the rate is estimated to a standard deviation of 0.25 mm/year. It is further assumed that the defect length grows in proportion with the depth. Using the method outlined in this paper, by calculating the uncertainty in the defect sizing for the future by adding the inspection uncertainty and the uncertainty in the corrosion rate, the allowable operating pressure decreases with time as given in Figure 5. All lines in Figure 5 have a target reliability level according to Safety Class Normal of 10^{-4} (annual failure probability). After 3-4 years the pipeline has to be de-rated from 155 bar if the reliability level shall be maintained, and after about 6 years the allowable pressure reduced to 100 bar. For the remaining lifetime the operating pressure is de-rated to 100 bar.

In order to further maintain the reliability level, the pressure can be further decreased, or a new inspection can be performed. In this example a new inspection is planned, and the operator has the option to inspect with a tool with the same accuracy as previously, or a very accurate inspection (Exact). For simplicity and to demonstrate the importance of the inspection uncertainty and the corrosion rate estimate uncertainty, in the example the inspection confirms that the defect has developed as the estimated mean value of the corrosion rate. The defect has after 6 years developed from 30% of t to 41% and length from 100 mm to 138 mm.

In Figure 5 the allowable pressure after the inspection at time 6 years is seen for an inspection with the similar accuracy as the first inspection. The decrease from 182 bar to 150 bar in allowable operating pressure is due to the growth of the defect, both having the same inspection accuracy. The mean corrosion rate is after the inspection estimated to be 0.3 mm/year, same as before, and confirmed by the inspection. If the same uncertainty as previous is still associated with the corrosion rate, the allowable pressure will decrease fairly fast, and a new inspection would have to be performed after another 3-4 years. But if the corrosion rate uncertainty based on the inspection findings etc. can be reduced from a standard deviation of 0.25 mm/year to 0.10 mm/year, the next inspection can be extended to 5-6 years.

If a very accurate (exact) inspection is carried out, as for instance external automated UT measurements, the sizing uncertainty can be minimal. The benefit in allowable pressure can be seen in Figure 5, but by time the allowable pressure decreases rapidly and the time to next inspection is about two years longer than for the less accurate inspection. However, this inspection, and especially with consecutive inspections, can with a higher degree of confidence form the corrosion rate and hence reduced the uncertainty in the rate estimate. In this case the time interval until the next inspection can be about 9 years.

An external automated UT inspection on a submarine pipeline can be costly, and cover only a fraction of the pipeline, but with a high degree of accuracy. An internal inspection with intelligent pig will cover virtually the full pipeline surface, but with less accuracy. A combination of these two inspection methods could be beneficial in many cases, as the external UT measurements can be used to confirm the internal inspection and may cover the most severe defects. In the example given above the most severe defect is considered, and if only this defect is inspected with an UT inspection other defects may have become critical and need also to be considered.

CONCLUSION

The Recommended Practice provides a means of relating allowable operating pressure to defect size together with its associated uncertainty, and can be used for inspection planning. This uncertainty depends both on the accuracy of the inspection and uncertainties in the corrosion rate estimate. The dominating effect of the uncertainties in the inspection and the development of defects is demonstrated. Probabilistic methods may be preferable in inspection planning since they are more flexible and provide more information, but the described method using RP-F101 can in many cases provide sufficient data for inspection planning.

REFERENCES

(1) DNV RP-F101, "Corroded Pipelines", 1999, DNV, Norway.

(2) DNV'96, "Rules for Submarine Pipeline Systems", 1996, DNV, Norway.

(3) DNV OS-F101, Offshore Standard, "Submarine Pipeline System", to be issued 1999, DNV, Norway.

(4) O.H. Bjørnøy, G. Sigurdsson, E.H. Cramer, B. Fu, and D. Ritchie, "Introduction to DNV RP-F101 Corroded Pipelines", OMAE 1999.

(5) G. Sigurdsson, E.H. Cramer, O.H. Bjørnøy, B. Fu, and D. Ritchie, "Background to DNV RP-F101 Corroded Pipelines", OMAE 1999.

(6) Ritchie, D., Voermans, C.W.M., Larsen, M.H., and Vranckx, W.R., "Probabilistic Tools for Planning of Inspection and Repair of Corroded Pipelines", OMAE98-0901, Lisboa, 1998.

(7) M. Marley, G. Sigurdsson, and O.H. Bjørnøy, "A Method for Assessing Corrosion in Pipelines and Planning For Intelligent Pigging", ELI Pipeline Rehabilitation and Maintenance Conference, Abu Dhabi, UAE, May 1999.

6 MAINTENANCE AND INSPECTION ACTIVITIES

6.1 SAFE OPERATING LIMITS & MAXIMUM PERMISSIBLE OPERATING PRESSURE

Prior to operating a pipeline it is essential to establish appropriate safe operating limits based on the pipeline design, test pressure, and any operational constraints that may need to be applied. The SOL for pressure will be related to the declared maximum permissible operating pressure (MPOP). The operator must ensure that the pipeline is maintained within these limits.

These SOL's, along with the declared MPOP, must be periodically reviewed and redeclared in order to confirm continued fitness for purpose.

6.2 PROVISION OF INFORMATION

The operator must provide those involved in operating, maintaining, inspecting, modifying and repairing a pipeline with sufficient written information as needed to perform their duties correctly and safely.

It is important that owners and occupiers of land through which a pipeline is routed, and third parties who may require to work in the vicinity, are reminded of, or made aware of, the pipeline's location.

Regular contact with owners and occupiers is essential (see 6.3 - Owner/Occupier Liaison).

In order to assist third parties the operator should make available pipeline location plans of a scale suitable to identify the approximate location of the pipeline. Use of 'one-call' systems may be beneficial in this respect.

It is further recommended that the pipeline route is marked with location markers which can also be used as an aid for aerial surveillance.

6.3 ROUTE SURVEILLANCE

It is essential to take all reasonable precautions to reduce the risk of pipelines being struck or damaged by third party activities. The risk of damage to a pipeline or pipeline system needs to be assessed in order to determine appropriate surveillance methods and frequencies.

Routine surveillance is employed to ensure that there are no unknown activities being carried out in the vicinity of pipelines that could cause damage to it. It should also be used to monitor building development that may result in proximity or population density infringements. Where infringements do occur they should be subjected to a risk assessment in order to determine if any additional protection or surveillance needs to be undertaken.

Where activities are being carried out by third parties in the vicinity of a pipeline, with the consent of the operator, it is necessary to monitor the activities and ensure agreed working methods and procedures are being adhered to.

The following are recommended methods of undertaking route surveillance.

Aerial Survey
A fast and efficient route survey can be carried out by helicopter. Typical activities that the observer should be looking for are :

- construction of any building work which may infringe proximity or population density criteria,
- any previously unknown third-party activity on or adjacent to the pipeline,
- the condition of pipeline hardware, such as marker posts.
- fires of any description - including straw burning,
- tree felling and timber transportation,
- discolouration of vegetation or other evidence of leakage,
- blasting or mineral extraction,
- ground movement,
- erosion and changing water courses,
- soil removal,
- tipping,
- overgrowth on easement.

Vantage Point Survey
As an alternative to the aerial survey, and for sections of pipeline not flown because of location, or due to the helicopter being unable to fly, vantage point surveys should be undertaken. This survey should cover the complete pipeline route, equivalent to an aerial survey. The surveyor will be looking for activities identified in the section on aerial survey when undertaking vantage point surveys..

Line Walk
A further alternative to aerial and vantage point survey would be a full line walk survey. The surveyor will be looking for activities identified in the section on aerial survey.

Owner/Occupier Liaison
Regular contact should be maintained with owners, tenants, occupiers, and managers of land through which a pipeline runs. Contact with local authorities, statutory bodies, and selective contractors is also recommended. The purpose of this is as a reminder of the existence of the pipeline, to establish future intended work in the vicinity of the pipeline, and as an aid to updating records of those responsible for the land.

Water Crossing Survey
Water crossings, and their survey techniques, can be divided into three categories :-

Major - greater than 50 metres deep, tidal and navigable.
These should be surveyed by remote operated vehicles or hydrographic survey.
Intermediate - 10-50 metres deep and navigable.
These should be surveyed using remote operated vehicles or divers.
Minor - Up to 10 metres deep.
These require a river bed and bank profile to be taken from a small boat.

Survey frequencies must be based on consideration of crossing depth, water flow rates, likelihood of erosion, undercutting etc.

All water course crossings that are wadeable need only be visually inspected as part of other pipeline external condition monitoring surveys.

Exposed Crossings

Exposed crossings should be inspected for security, mechanical defects, condition of supports, paintwork and wrapping. The frequency of inspection must be determined by the location of the crossing and its vulnerability to external interference, including vandalism.

Nitrogen Filled Sleeve Monitoring

Nitrogen filled sleeves should maintain a positive nitrogen pressure. This pressure should be checked regularly. For sleeves not holding their charge more frequent checks must be made in order to top up to ensure a positive pressure is maintained.

6.4 CONDITION MONITORING

The integrity of a pipeline is monitored periodically using on-line inspection techniques which can be either internal or external in nature, or alternatively by hydrostatic testing. Condition monitoring establishes a continued fitness for purpose assessment of pipelines.

Internal Inspection

Where practical, pipelines should be constructed to allow internal inspection utilising on-line inspection tools. A series of preparatory pigging runs, including cleaning, geometric, and profile are undertaken to ensure safe passage of the inspection pig. This inspection employs a magnetic flux leakage, or ultrasonic pig to detect metal loss in pipe walls and will locate and size areas of corrosion and mechanical damage. The magnetic flux leakage pig can also detect circumferential cracks.

For pipelines which may be subject to stress corrosion cracking consideration should be given to the use of an elastic wave or transverse flux leakage internal inspection tool.

External Inspection

Where the use of an internal on-line inspection tool is not practical then an external above ground survey should be undertaken. This should involve a corrosion protection close interval potential survey (CIPS) and a coating defect survey (Pearson).

Close Interval Potential Survey (CIPS)

The CIPS survey determines the actual level of cathodic protection being experienced along the pipeline by measuring the pipe to soil potential and hence corrosion protection levels.

Coating Defect Survey (Pearson Survey)

This survey is carried out to determine the soundness of pipeline coating by looking for defects. It measures an impressed electrical signal along the pipe.

Hydrostatic Testing

For pipelines not suitable for internal or external on-line inspection hydrostatic testing should be undertaken. With certain liquids this test can be carried out by utilising that product.

A frequency strategy for undertaking appropriate condition monitoring techniques described above must be developed taking into account legislative requirements, operation duty, corrosion and other evidence of deterioration, and pipeline life requirements.

6.5 CATHODIC PROTECTION

All pipelines should be protected from corrosion by protective coatings and cathodic protection systems. The cathodic protection systems can be of sacrificial anode type or impressed current type. The selected system must be monitored to ensure continued correct operation. The system should also be applied to appropriate steel sleeves.

A sacrificial anode system, normally utilising magnesium anodes, should be monitored regularly at pipeline locations of known low protection followed by a routine full monitoring inspection.

An impressed current system should be monitored regularly at the point of application of the protective current (transformer/rectifier units) followed by a routine full monitoring inspection.

6.6 REPORTING OF DAMAGE & DEFECTS

All damage to pipelines must be recorded and reported to the pipeline operator. Any defect found, generally following a maintenance or inspection activity, must also be recorded and reported. Following the report, an assessment of the damage or defect must be carried out by a competent person. The results of the assessment and of any remedial actions must also be recorded.

It is recommended that the operator establish a database of all reported damage and defects, which can be used to assist in establishing continued fitness for purpose.

Following damage to a pipeline it may be necessary to reduce the safe operating limits and maximum permissible operating pressure until remedial actions have been carried out.

6.7 EMERGENCY MAINTENANCE AND REPAIR

In order to respond to maintenance and repair requirements following pipeline damage, the operator should have immediate access to appropriate plant, materials and expertise as is needed to undertake remedial action (see 8 - Emergency and Repair).

6.8 DEFECT ASSESSMENT

The operator should have in place a process for assessing defects. This should include a competent person or persons with suitable qualifications and experience to be able to assess the effect the defect has on the integrity of the pipeline against appropriate defect acceptance criteria. If the defect is beyond the acceptance level the competent person should recommend repair methods or other remedial actions to re-establish fitness for purpose. (see 7 - Defect Assessment).

6.9 MODIFICATION AND REPAIR

The operator should have in place a system for recording design details of all proposed modifications and repairs and having them appraised and approved prior to being carried out. (see 10 - Modification and Repair Procedure).

6.10 RECORDS
The operator should have in place a system for generating and maintaining sufficient records as to enable continued fitness for purpose to be demonstrated. (see 9 - Keeping of Records).

7 DEFECT ASSESSMENT

There are no generally accepted international codes for the assessment of damage and pipeline repair. The ANSI/ASME B31.G Manual for Determining the Remaining Strength of Corroded Pipelines give guidance on corrosion damage whilst Transco has extended these principles to cover assessment of mechanical damage as well as corrosion in their P11 standard - Procedures for Inspection and Repairs of Damaged Steel Pipelines.

8 EMERGENCY AND REPAIR

Regardless of the robustness of the maintenance and inspection regime unpredicted pipeline failures do occur. It is therefore essential to react and respond quickly for both safety and economic reasons.

Emergency procedures must be in place to react to pipeline 'hits', from assessing the damage, categorising damage then carrying out repairs as necessary, to isolating the effected section and then carrying out repairs.

The emergency response must also include the availability of plant and materials, equipment needed to make safe and work on pipelines underpressure, together with trained teams, who can react quickly to make the situation safe and undertake permant repairs.

Damage and defects must be repaired to a standard that will maintain the overall fitness for purpose level of the pipeline.

9 KEEPING OF RECORDS

It is essential that a complete and accurate set of records are produced and retained which identify the pipeline's route, construction details, modifications, repairs, material certificates, test records etc., in able to confirm that the pipeline is suitable for operating within its declared safe operating limits and maximum permissible operating pressure.

10 MODIFICATION AND REPAIR PROCEDURE

All diversions, modifications, repairs etc. should be subject to a procedure that checks, appraises and approves the design of the proposed work and has it authorised by the user. The system must incorporate the collation of records to ensure continued fitness for purpose.

11 REVIEW OF STRATEGY

A process to continually review all maintenance and inspection practices should be established. This should include recording and keeping data for analysis of existing maintenance and inspection activities and frequencies, damage, defects, repairs and modifications, which can be used to continually develop the most safe, reliable and cost effective maintenance and inspection strategy.

Table 1

STATUTORY REQUIREMENTS

	MAINTENANCE AND INSPECTION ACTIVITY	REGULATIONS PSTGCR	PSR	
6.	Maintenance & Inspection	12	13	General maintenance requirements
6.1	Safe Operating Limits	7	6,11	S.O.L. pressure is determined from MPOP declaration.
6.2	Provision of Information	5,11	16	To enable operatives to carry out duties and to provide owners/occupiers and third parties with relevant
6.3	Route Surveillance	-	13,15	To maintain control of third party activities.
6.4	Condition monitoring	9,12	13	Internal, CIPS, Pearson, External & hydrostatic testing.
6.5	Cathodic Protection	12	13	Included as maintenance requirements.
6.6	Reporting of Damage and Defects	10	15	Reporting requirements.
6.7	Emergency Maintenance & Repair	10	12	General requirements.
6.8	Defect Assessment	4,10	12	General requirements.
6.9	Modification & Repair	4	5,10,22	Recording, appraisal, approval requirements.
6.10	Records	8,13	23	Record requirements.
7.	Emergency and Repair	10	23,24,25	Emergency procedures.

Table 1

STATUTORY REQUIREMENTS

	MAINTENANCE AND INSPECTION ACTIVITY	REGULATIONS		
		PSTGCR	PSR	
6.	Maintenance & Inspection	12	13	General maintenance requirements
6.1	Safe Operating Limits	7	6,11	S.O.L. pressure is determined from MPOP declaration.
6.2	Provision of Information	5,11	16	To enable operatives to carry out duties and to provide owners/occupiers and third parties with relevant information.
6.3	Route Surveillance	-	13,15	To maintain control of third party activities.
6.4	Condition monitoring	9,12	13	Internal, CIPS, Pearson, External & hydrostatic testing.
6.5	Cathodic Protection	12	13	Included as maintenance requirements.
6.6	Reporting of Damage and Defects	10	15	Reporting requirements.
6.7	Emergency Maintenance & Repair	10	12	General requirements.
6.8	Defect Assessment	4,10	12	General requirements.
6.9	Modification & Repair	4	5,10,22	Recording, appraisal, approval requirements.
6.10	Records	8,13	23	Record requirements.
7.	Emergency and Repair	10	23,24,25	Emergency procedures.

Table 2

PIPELINE MAINTENANCE AND INSPECTION MATRIX

	Drivers / Activity	2.1 LEGISLATIVE COMPLIANCE PSTGCR	PSR	2.2 OPERATIONAL DUTY	2.3 BUSINESS AND ECONOMIC FACTORS	2.4 SAFETY AND ENVIRONMENTAL ISSUES	2.5 POTENTIAL DAMAGE MECHANISMS	2.6 THIRD PARTY ACTIVITIES
6.1	Safe Operating Limits	7	6,11	Based on Operating Criteria	Maximises Utilisation	Defines Overpressure Limitations	Prevents Failure and Construction Defects	-
6.2	Provision of Information	5,11	16	Assists in Maintaining Integrity	Minimises Interference	Maintains Integrity and Identifies Land Use	Minimises Interference	Minimises Interference
6.3	Route Surveillance	-	13,15	Reduces 3rd Party Activities	Minimises Interference	Monitors Land Use & Ground Conditions	Minimises Interference	Minimises Interference
6.4	Condition Monitoring	9,12	13	Maintains Asset Life	Maximises Availability	Monitors Land Use & Ground Conditions	Detects Impact, Corrosion and Coating Damage	Detects Mechanical Damage
6.5	Cathodic Protection	12	13	Maintains Asset Life	Maintains Integrity	Maintains Integrity	Reduces Corrosion Damage	Maintains Integrity
6.6	Reporting of Damage & Defects	10	15	Reduces Unexpected Failures	Review Maint & Insp Strategy	Review External Factors	Identifies Problems	Must Report Damage
6.7	Emergency Maintenance & Repair	10	12	Reduces Downtime	Reduces Downtime	Quick Response & Product Containment	Quick Remedial Action	Quick Remedial Action
6.8	Defect Assessments	4,10	12	Sets Criticality Limits	Review Maint & Insp Strategy	Ensures Integrity	Categorise and Assess Defect Significance	Assess Damage Significance
6.9	Modification & Repair	4	5,10,23	Ensures Integrity is Maintained	Review Maint & Insp Strategy	Maintains Integrity	Controls Risk	Reinstates Integrity
6.10	Records	8,13	23	Provides f.f.p. Evidence	Provides continued f.f.p. Evidence	Demonstrates Integrity	Monitors Risk	Identify Location

High Pressure: Uprating, Repair, Revalidation, and Decommissioning

C571/003/99

Fatigue and seismic assessment of hydro-electric pipelines

G J GIBBERD and **B M ADAMS**
WS Atkins Consultants Limited, Almondsbury, UK
M SEATON
Scottish and Southern Energy plc, UK

SYNOPSIS

Advanced analytical techniques are now commonplace in a variety of industries, for example non-linear finite element analysis and computational fluid dynamics (CFD) are widely used in the design of anything from cars to prosthetic heart valves. The application of such techniques to the management and operation of ageing plant items such as pipelines is, however, less commonplace. This paper presents a case study in which a range of high technology techniques, adapted from a variety of industries, were applied to a 70 year pipeline system owned and operated by Scottish and Southern Energy (SSE). Through the use of techniques such as elasto-plastic finite element analysis, high cycle fatigue analysis and crack growth assessments, combined with measured plant data, WS Atkins were able to define a cost-effective asset management strategy for the Rannoch pipelines.

1 INTRODUCTION

Scottish & Southern Energy plc (SSE) is the largest generator of conventional hydro power in the UK.. The company owns and operates 66 hydro power stations in the Scottish Highlands and Islands with a total installed capacity of 1060 MW, as well as the 300MW Foyers pumped storage scheme. The majority of the hydro schemes were constructed during the intensive post-war hydro-electric development between 1947 and 1967. However, Rannoch Power Station, along with others in the Tummel Scheme, was constructed by the private Grampian Electricity Company during the 1920s and 1930s.

As part of a major asset management strategy, SSE are currently undertaking a rolling programme of refurbishment to many of these stations. The refurbishment programme is centred on ensuring safe, efficient and economic running of their stations for several decades into the next millennium. A major aspect of the upgrade activity involves replacement of the original turbines with modern designs of increased efficiency. At present there are no plans to replace the original above ground pipelines, however SSE are undertaking a series of assessments to identify what modifications, if any, are required to ensure the structural integrity of these ageing pipelines for their foreseeable future life.

As part of the refurbishment programme, WS Atkins were recently commissioned by SSE to perform a comprehensive study of pipeline structural integrity for Rannoch Power Station. The station is located on the north shore of Loch Rannoch, approximately 13km west of Kinloch Rannoch, and operates on the waters of Loch Ericht. During commissioning of Rannoch in the early 1930's, the pipelines were observed to vibrate significantly when the turbines were operated at full capacity. As a result of this the turbines have operated under a load cap for their entire lifetime to prevent excessive fatigue damage to the pipelines.

The objectives of the study were threefold:

- Diagnose the cause of pipeline vibrations and determine the implication of the vibrations on the future operations of the replacement turbines.
- Investigate the fatigue life usage of the pipelines to date, and determine whether any modifications are required to ensure adequate fatigue life for the station's foreseeable future usage.
- Assess the integrity of the pipelines against the 10^{-4} per anum seismic event defined for the site, anchored to a ground acceleration of 0.2g.

SSE are currently required to assess the seismic risk to their dams and consider it a logical extension of these requirements to include the risk to large pipelines where significant free discharge could occur in the event of a gross failure.

2 DESCRIPTION OF RANNOCH POWER STATION

Rannoch Power Station comprises three vertical axis Francis turbines, housed in a turbine building on the north shore of Loch Rannoch. The original turbines were rated at 16MW each, but due to the pipeline vibration problems, the station was operated with a total output limit of 42MW. The new, uprated turbines were installed in early 1999.

The turbines are fed with water from Loch Ericht, some 6km away. A single, concrete-lined low pressure tunnel runs from Loch Ericht to approximately 1km from the turbine house. At this point the tunnel bifurcates to form two steel-lined tunnel sections. These steel-lined tunnels bifurcate at the valve house, resulting in four pipelines. One of these is immediately blanked off at the valve house, whilst the remaining three run down above ground to the turbine building. Figure 1 shows the pipelines viewed from the valve house, looking down towards Loch Rannoch and the turbine building.

The three pipelines are nominally identical and have a 95in (2.4m) diameter. Along their length, five concrete anchor blocks are used to restrain the pipelines at changes of gradient. Each pipeline is constructed in 23ft (7m) long sections using longitudinally butt-welded mild steel plates. Circumferential spigotted and riveted joints are used to connect pipeline sections. Typical for its time, the Rannoch pipelines were designed primarily against the internal pressure head, and hence the pipe wall thickness varies along the length, rendering the pipe walls quite flexible in the upper low-head regions. At the valve house a thickness of ½ " (12.7mm) is used and at the turbine building the thickness is $^{15}/_{16}$" (23.8mm).

In between the concrete anchor blocks, and at approximately 48' (14.6m) intervals, the pipelines are supported on mild steel saddles. The saddles consist of angle sections riveted to a web plate and a saddle plate. The saddle plates are fillet welded to the pipelines all around the 150° arc of support, including along the top of the horns. Each saddle sits on a concrete plinth, which is keyed into the bedrock. To accommodate longitudinal thermal expansion of the pipelines, four rollers sit between the saddle base plate and the plinth. The rollers are keyed into the plinth to provide a degree of lateral restraint. Figures 2 shows a typical support saddle.

3 DIAGNOSIS OF VIBRATION PROBLEM

As discussed previously, Rannoch has operated under a 42MW output limit since commissioning, due to vibration of the pipelines when the turbines are run at high load. During early 1998, prior to the installation of the new turbines, WS Atkins conducted a study to diagnose the cause of the vibrations and determine whether any modifications would be required to prevent the new turbines from having to be load limited.

The diagnostic study was based on site measurements made by WS Atkins engineers over a three day period in January 1998. The pipelines were instrumented externally with accelerometers and internally with pressure transducers. A detailed test programme was designed such that the following issues could be investigated:

- The hydraulic resonant frequencies inherent in the pipelines.
- The mechanical resonant frequencies of the pipelines.
- Any correlation between pressure fluctuations and mechanical vibration.
- Any correlation between station power output and mechanical vibration.

In the light of the measurement programme, the validity of a number of potential vibration mechanisms was assessed. Phenomena such as vortex shedding from valve bodies, turbulence generation at pipe bifurcations and cavitation emanating from debris and drainage sumps were considered. However, it was concluded that the pipe vibrations were the manifestation of an internal hydraulic resonance of the entire pipeline system, driven by a hydraulic instability at the turbines, which in turn was forcing a structural response. An explanation of this mechanism, and the key evidence supporting its validity for the Rannoch pipelines is provided below.

Under normal operation Francis turbines generate pressure oscillations from a number of sources. A good review of the phenomena is given in [1]. Since they are reaction type machines, these pressure oscillations propagate upstream via the normal waterhammer process. A key driver of pressure oscillations is often the well known draft tube vortex of the Francis turbine. This is present to some extent under all load conditions, and leads to surging within the draft tube. The frequency of the vortex is typically 30-40% of the runner rotational speed, depending on the turbine loading. For the 428rev/min Rannoch turbines, the draft tube vortex shedding frequency would be expected to be around 2 to 3 Hz.

During the monitoring programme, power spectral density information was derived from pressure transducer and accelerometer measurements taken during a test in which one of the pipelines was experiencing significant vibrations. The pressure transducer spectra showed a clear resonant peak at 2.5Hz indicating a hydraulic resonance in which a standing pressure wave is set up in the pipeline. Given that the Rannoch turbines have a draft tube vortex shedding frequency of 2 to 3 Hz, it is reasonable to conclude that the hydraulic resonance is being excited by draft tube vortices.

The accelerometer spectra also showed a clear peak in the pipeline's structural response at 2.5Hz. From simple mass and stiffness considerations, it is known that the pipelines do not have any structural resonant frequencies at or around 2.5Hz, and consequently it is reasonable to conclude that the structural vibrations are being forced by the resonant pressure fluctuations. The forced response produces a "panting" motion with the pipelines repetitively inflating and relaxing. At the saddle locations, the "panting" motion becomes an ovalling motion due to the radial restraint provided by the saddle. This mechanism has been observed in other hydro-electric stations, such as Magnox Electric's Maentwrog station in North Wales [2]. In the case of Maentwrog, the vibrations were sufficient to require modifications to the draft tube and support saddles to ameliorate pipeline high cycle fatigue life usage. There are many reported cases of pipelines cracking in the region of the saddle horns from similar mechanisms.

Having identified the turbines as the source of pipeline vibration, consideration was given to what effect the new turbines would have on the level of vibration. Since the refurbishment will not alter the hydraulic characteristics of the pipelines, and since the new turbines will have a similar range of draft tube vortex frequencies, it was concluded that the pipelines will still be subject to forcing from hydraulic standing waves. The magnitude of the pressure oscillations, however, are likely to be reduced because the new turbines will have a more efficient draft tube profile and will be fitted with provision for air injection to suppress surging. It was therefore concluded that the new turbines do not present an increased risk to the pipelines from vibration fatigue. However, in view of experiences at the Maentwrog station where the installation of new turbines increased the levels of pipeline vibration, it was considered prudent to perform a fatigue assessment to confirm that sufficient fatigue life is available to accommodate the pipelines future usage.

4 FATIGUE ASSESSMENT

Following on from recommendations made during the diagnostic study, WS Atkins undertook a detailed fatigue assessment of the Rannoch pipelines. The areas of pipeline at most risk from fatigue damage are around the support saddle locations. The regions just above the saddle horns are of particular concern, since the sudden discontinuity at these points can give rise to significant circumferential bending stresses, and under vibrational loading high cycle fatigue therefore becomes a potential threat. The pipelines would have originally been designed against the static pressure and gravity loads only, and no consideration would have been given to dynamic stresses; indeed even modern codes give little guidance in this aspect.

A fatigue assessment of the Rannoch pipelines is subject to two fundamental difficulties:

- The stress field around the saddle horns is extremely complex. Using code based arguments for pressure vessels supported on saddles [3], localised stresses in excess of yield are predicted under static loads alone. Consequently, the behaviour under dynamic loading can only be accurately established by considering the elasto-plastic material properties. (Note that the peak static stresses are highly localised and predominantly secondary in nature; local plasticity is thus tolerable without affecting the global integrity structure).

- The predominant frequency of vibration is known to be 2.5Hz and consequently there is the potential for approximately $8x10^9$ cycles to accumulate over a 100-year lifetime. Reliable material fatigue data for this number of cycles is rare, with most modern design codes only presenting data up to around 10^8 cycles.

As a result of these issues, it was not possible to perform a classical, design code based fatigue assessment. Consequently a hybrid methodology was developed, using finite element analysis to calculate stress ranges and material data adapted from the nuclear industry to determine fatigue damage.

The LS-DYNA3D finite element package was used to model the pipelines. LS-DYNA3D is used extensively in the automotive, rail and aerospace industries to model structural non-linear behaviour during transient events such as crashes or impacts, and was chosen here as it offered the facility to model plastic strain cycles. Figure 3 shows the LS-DYNA3D model of a pipeline section close to the valve house, when at its nominal operating pressure of 4.5barg. The displaced shape has been magnified to highlight the ovalling shape over the support saddle, and the contours presented are based on the von Mises stress on the inner surface. From Figure 3, the peak stress at the saddle horns is predicted to be around 235MPa. This compares favourably with a value of 240MPa predicted using the code based methods of [3].

To gain a better understanding of the pipeline stresses, a transient analysis was performed to simulate the behaviour of the pipelines when being filled with water and pressurised, followed by depressurisation and emptying. Such watering and de-watering cycles would have occurred during original commissioning and subsequently when maintenance or inspection procedures have required man access to the pipelines. They are significant because they are the largest load cycle that the pipelines experience. Figure 4 presents the results of the analysis in terms of inner and outer fibre circumferential stress at the saddle horn as a function of internal pressure. The start of the watering phase and the end of the dewatering phase correspond to the situation where the pipeline is full of water but not pressurised.

There are a number of interesting features to note from Figure 4:

- The initial filling of the pipelines causes tensile yielding on the inner surface and compressive yielding on the outer surface.
- Due to pressure stiffening effects, increasing the water pressure tends to relieve and then reverse the stresses at the saddle horns. At the normal operational pressure for this section of pipeline, 4.5barg, the outer surface is approaching tensile yield whilst the inner surface is in elastic compression.

- Based on the above observation it can be concluded that any fatigue cracks are most likely to initiate on the outer surface and grow towards the inner surface.
- Periodic watering and dewatering of the pipelines will cause plastic strain cycling of the outer surface around the saddle horns. The maximum local strain range predicted by LS-DYNA3D is in the order of 2%, which when combined with the dewatering interval of 6 years, does not pose a low cycle fatigue threat to the structural integrity of the pipelines.

The results presented in Figure 4 were used as the basis of a high cycle fatigue analysis for the outer surface of the pipelines, adjacent to the saddle horns. Using the measurements of pressure fluctuation taken during the diagnostic study, Figure 4 was used to provide corresponding stress ranges. The fatigue damage associated with the calculated stress cycles was assessed using material fatigue data taken from the nuclear industry [2]. The data presented in [2] are used extensively in the assessment of nuclear power plant subject to large numbers of cycles, and provide fatigue life information for a range of classes of welded features. Despite the age of the pipes (~70years), the fatigue data were considered to be applicable to the type of ferritic/pearlitic steel used. A classification appropriate for low grade fillet welds between two plates in shear was used for the Rannoch assessment, since on-site inspections had shown the saddle to pipeline weld to be locally cracked with significant undercutting of the parent metal around the weld. Figure 5 presents the results from the fatigue assessment, showing the proportion of fatigue life used per year as a function of mean operating head and pressure oscillation magnitude, since uncertainty in these factors was identified as having most significant influence on the fatigue results.

The results presented in Figure 5 assume that the magnitude of the pressure fluctuations remain constant throughout the entire period of operation. In reality, the turbine driven hydraulic resonance causing the pressure fluctuations is a complex, unstable phenomenon influenced by a large number of disparate factors such as water chemistry, control system backlash, temperature, flow rate etc. During the measurement programme, pipelines were observed to drift in and out of resonance even whilst the turbine load settings were held nominally constant. Unfortunately long term measurements of the pressure fluctuation amplitudes and the periods of time for which they occur do not exist, and hence simplifying assumptions were used to provide a pragmatic means of gaining an upper bound to the pipelines' fatigue damage assessment.

Assuming a pipeline lifetime requirement of 100 years (70 years operation to date plus 30 years following refurbishment), Figure 5 shows that pressure fluctuations in the range 0.12barg to 0.09barg RMS can be tolerated, dependant on the mean pressure head. However, during the diagnostic test programme, pressure oscillations in the range 0.10barg to 0.15barg were measured during periods of resonance. It is interesting to note how low the pressure fluctuation required is to cause significant vibration: a reflection of the insufficient attention paid to wall stiffness in low-head regions. Also, in many instances, these magnitudes were achieved when the turbines were operating below the load limit of 42MW; the load limit has not therefore been sufficient to eliminate vibration, and the corollary is that some fatigue damage will have been accrued over the 70 years life to date.

Based on experience from the diagnostic test programme, it was considered likely that the pipelines would experience hydraulic resonant conditions for only short periods of time. However, without any long-term data to back this up, it was not possible to use the Figure 5 results to provide a more accurate summation of the fatigue damage. Consequently the upper bound analysis of Figure 5 was not sufficient alone to provide a robust fatigue case for the pipelines.

The next stage of the integrity assessment was to carry out a defect tolerance and crack growth assessment to determine whether leak-before-break arguments and/or regular inspection could be used in conjunction with the fatigue analysis to underwrite long term operation of the pipelines. The analysis started from the understanding that the area of concern is cracks initiating in the region of dynamic tensile stress on the outer surface of the pipes local to the saddle horns, as was indicated by the LS-DYNA3D modelling of the pipelines. The assessment used the PD6493 [4] procedure, and an established method as adopted by the nuclear industry. The results of the assessment predicted that a fully penetrating defect would become critical at a length of just 50mm. Based on this result it was concluded that a leak-before-break argument could not be substantiated. Further analysis demonstrated that a defect developing from the outer surface of the pipeline towards the inner surface would grow 5mm over a four year period before arresting due to the compressive stress regime on the inner surface. This provided a reasonable level of confidence in the long term integrity of the pipelines, however it was considered that future operation should be underwritten by a scheduled programme of non destructive testing (NDT).

Using the LS-DYNA3D models for various saddle positions, WS Atkins were able to determine the most critical saddle locations and hence provide recommendations for a tailored NDT programme. Based on the predicted crack growth rates, a programme of 5-yearly NDT surveys was recommended for just 39 of the 162 saddles (those in the uppermost thin-walled region). Furthermore, the insight gained from the LS-DYNA3D modelling enabled the outer surfaces around the saddle horn to be highlighted as the only areas requiring NDT. This represents a major reduction in the cost of the NDT surveys, since the pipelines do not require de-watering and the surveys can be done whilst the station is operational. The cost of the recommended NDT programme is estimated to be an order of magnitude smaller than would be required to modify the support saddles and two orders of magnitude lower than would be required to replace the pipelines. This provides an indication of the value of using high technology assessment techniques as part of an asset management strategy for ageing pipelines.

5 SEISMIC ASSESSMENT

Concerns about the seismic safety of concrete dams have been growing during recent years, partly because the population at risk downstream of major dams continues to expand and also because it is increasingly evident that the seismic design concepts used for existing dams were inadequate. A logical extension to these concerns is to include the risk to large surface pipelines, where free discharges could result in a significant release of water giving rise to an unacceptable level of risk to people and property. The seismic case for pipelines has to date been based on a qualitative view of the tolerability of consequences of failure (e.g. they are located in remote areas away from centres of population, and the time at which any people would be at risk from flooding is sufficiently small to be discounted). These arguments have recently been reviewed in the light of factors such as increased land access rights for leisure activities, and consequently SSE now desire a more deterministic case to be made.

In parallel with the fatigue assessment described above, WS Atkins undertook a seismic assessment of the Rannoch pipelines. It is understood to be the first hydro-electric pipeline of its kind to be assessed against seismic loading, and the results obtained provide generic information about the seismic behaviour of pipelines, which may be applicable to other pipelines. In the event of an earthquake, pipelines such as the Rannoch pipelines are vulnerable to a variety of potential failure modes; ground motion may destabilise and collapse the support structure (i.e. saddles and anchor blocks), bending or rupture of the pipe may arise due to inertia loading, and differential movement of pipe supports may overstress the pipe material and cause joints to fail. The main issue is that pipelines are rarely if ever designed for significant lateral loads, and any allowances made for any wind loading will be small in comparison with seismic inertial loads.

The assessment was carried out against the 1 in 10,000 year return earthquake (the 10^{-4} seismic event), as used previously for the seismic assessment of Ericht Dam. (The use of the same seismic levels as at the dam was considered appropriate given the dam's relative proximity and the fact that the pipelines are supported on the local rock). This represents a severe earthquake with horizontal ground accelerations in the order of 0.2g. The simplest and most common means of seismically assessing structures and equipment is to use the secondary response spectrum approach. The secondary response spectrum shows the peak acceleration that would be experienced during the seismic event by a floor mounted single degree of freedom system, plotted as a function of the system's natural frequency. Figure 6 shows the secondary response spectrum for the 10^{-4} event.

Using the secondary response spectrum method, the structure is simplified to a single degree of freedom system, and its natural frequency determined. The peak seismic acceleration is then extracted from the secondary response spectrum at the appropriate frequency, and applied as a static inertia load to the structure. To apply this method to the Rannoch pipelines it is necessary to assume that the support saddles are infinitely stiff and that the pipeline acts as a multispan beam between encastre supports at the anchor blocks. Using this approach, the fundamental bending frequency of the pipeline system is calculated as 12Hz and seismic accelerations of 0.6g are predicted from the secondary response spectrum. Under these accelerations, gross failure of the pipelines is predicted at the saddle locations.

In the case of the Rannoch pipelines, the secondary response approach is considered to be overly conservative for the following reasons:

- It does not take account of the vibration mode shapes, which in the case of a multi-span pipe such as the pipelines, will act to reduce the inertia loads experienced at each saddle location.

- The support saddles have some inherent lateral flexibility, and hence the assumption that the pipeline behaves as a single degree of freedom system is a gross simplification.

- The support saddles are susceptible to sliding and rocking under lateral loads. This introduces significant non-linearity into the seismic response of the pipelines, which invalidates the assumption of a linear-elastic single degree of freedom system.

To address these issues and provide a more realistic, less conservative representation of the pipelines' seismic integrity, a finite element model of a pipeline section between anchor blocks was constructed using the ABAQUS package. ABAQUS is a well known finite element package and is used extensively in a variety of industries for modelling the dynamic behaviour of structures and mechanisms.

Within the ABAQUS model, simple linear beam elements were used to represent the pipeline sections between support stools, enabling the bending and torsional response to be modelled. Linear elastic spring elements were used to model the lateral stiffness of the support saddles. The spring stiffnesses were estimated using the LS-DYNA3D model developed for the fatigue assessment. Figures 7a and 7b provide diagrammatic representation of the ABAQUS model showing two typical vibrational mode shapes. The 5.7Hz vibrational mode shown is characterised mainly by lateral "swaying" of the support saddle arms whilst the 12Hz mode is characterised by longitudinal flexure of the pipelines themselves.

The seismic behaviour of the pipelines was investigated by applying displacement time histories to the support saddle and anchor block bases within the ABAQUS model. The displacement time histories were generated using a proprietary computer package, and were constructed such that they had the same frequency content as the secondary response spectra shown in Figure 6. Throughout the modelled seismic event, restraint forces and moments were extracted from a variety of locations on the pipelines.

A qualitative review of the pipelines and support structure design was carried out to determine potential failure modes under a seismic event. The key areas of concern were found to be:

- The shear keys between the saddle roller bearings and the concrete plinths provide the only resistance to lateral loading on the support saddles. If these keys fail, there is potential for the support saddles to slide.
- Rocking of the saddles. Since there is no positive connection between the support saddles bases and the roller bearings, the saddles may rock under lateral loading. If the rocking amplitude is sufficiently large it could cause the shear keys to become disengaged and hence the saddles to slide. Rocking can also lead to impact loads between the saddles and the saddle base plates.
- Overstressing of the pipelines could occur at the anchor blocks since the pipes are encastre and therefore directly excited by the ground motion. The forces at the anchor blocks would also be increased by lateral flexure and twisting of the pipe due to saddle rocking or lateral sliding.
- At the support saddle locations, the pipelines could rupture due to either bending or buckling collapse of the pipe walls, or from the saddles "punching" their way through the thin shell wall.

Using a combination of hand calculations and results from the LS-DYNA3D model, quantitative criteria were derived for the failures described above. These criteria were expressed in terms of restraint forces and moments and could hence be readily compared with the time history output from the ABAQUS model. To illustrate this method, Figure 8 shows saddle restraint force output from one of the ABAQUS runs, plotted against the seismic failure criteria for rocking of the saddles.

A number of sensitivity studies were carried out to bound uncertainties in aspects such as the flexibility of the support saddles and the phase difference between ground motions felt at different locations along the pipelines.

From the various ABAQUS runs, the following conclusions were drawn:

- There will be some localised yielding of the pipelines at the support saddles and anchor blocks, however the maximum plastic strain is an order of magnitude less than that required to cause rupture.

- There will be some lateral rocking of the support saddles, but not sufficient to cause gross instability of the support structure.

- In some instances, the shear keys that prevent lateral movement of the support saddles will fail, however the accumulated lateral displacement at each such saddle will not be sufficient to threaten the integrity of the pipelines through loss of vertical support.

From these results it was concluded that a safety case could be made against the 10^{-4} seismic event and that modifications are not required to either the pipelines or the supporting structure. As with the conclusions from the fatigue assessment, this represents a significant cost saving to SSE since seismic modifications to either the pipelines or the support structures would represent a significant undertaking.

6 CONCLUSIONS

The case study presented here has demonstrated how advanced engineering analysis techniques can be successfully applied to the asset management of large pipelines.

In the case of dynamic loading, high-cycle fatigue can be a threat, and it has been shown how detailed numerical stress modelling can be synthesised with test data to investigate the long-term integrity of pipeline structures.

The techniques presented are similarly applicable to the integrity and remnant life assessment of many types of pipework and support structures, and offer a cost-effective route to defining plant management strategies.

REFERENCES

[1] Grein, H, 'Vibration phenomena in Francis turbines: Their causes and prevention'. Escher Wyss paper, Tokyo Symposium, 1980.

[2] Gibberd, GJ, 'Severe hydro-mechanical vibrations following the upgrading of a 30MW UK hydro station – a case study'. International Water Power and Dam Construction Magazine Conference - Uprating and Refurbishing Hydro Plants, Montreal 1 – 3 October 1997.

[3] British Standards Institution, 'Specification for unfired fusion-welded pressure vessels: BS5500, 1997'
[4] British Standards Institution, 'Guidance on methods for assessing the acceptability of flaws in fusion welded structures: PD6493, 1991'

Figure 1 : View of the Rannoch Penstocks

Figure 2 : View of a Support Saddle

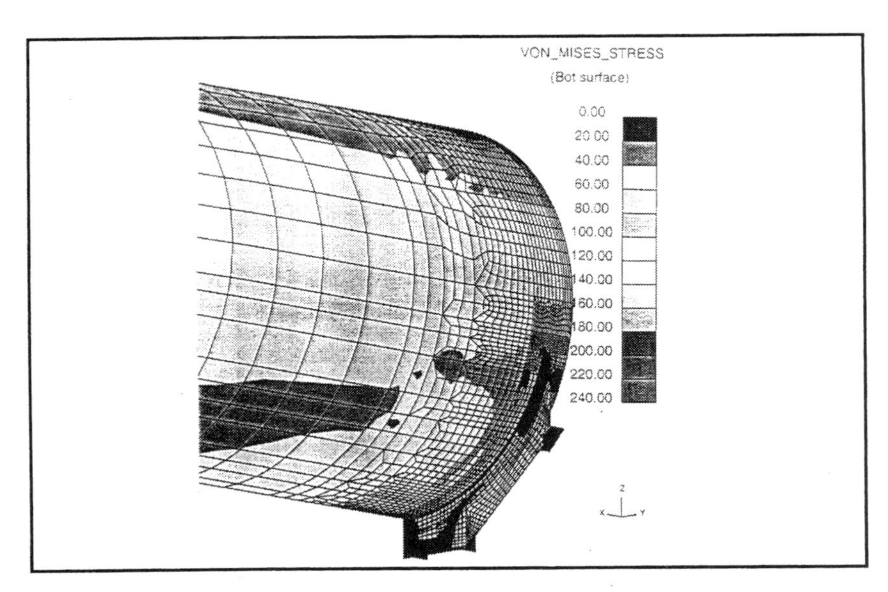

Figure 3 : LS-DYNA3D Model of the Penstock and Saddle

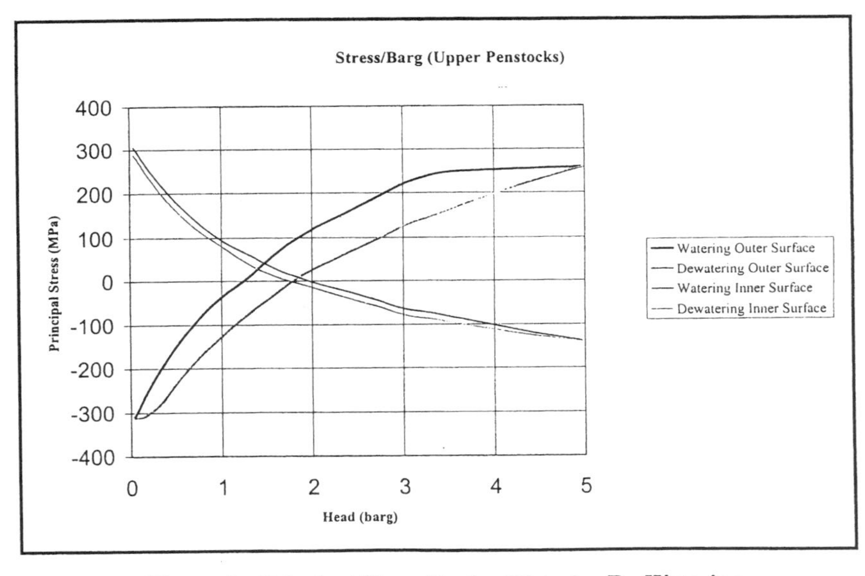

Figure 4 : Principal Stress During Watering/De-Watering

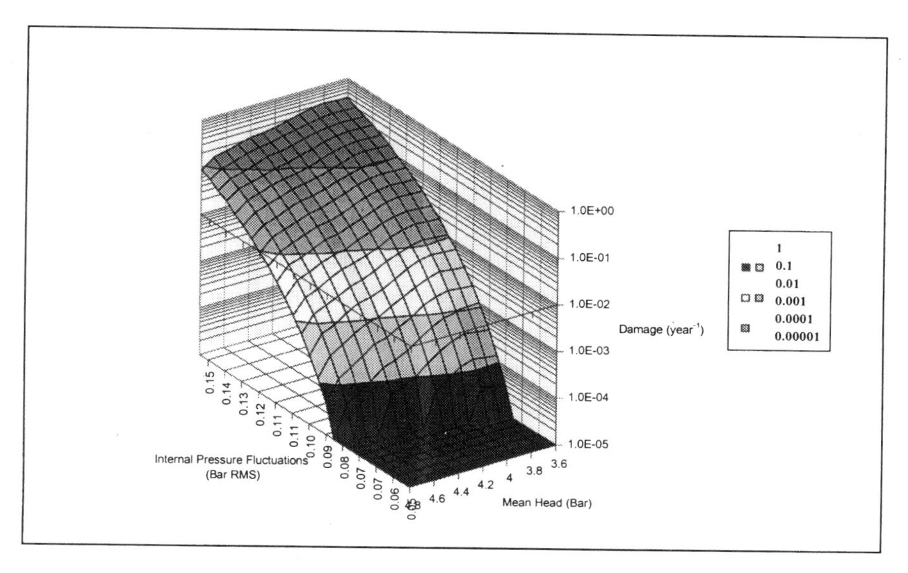

Figure 5 : Results of the fatigue Analysis

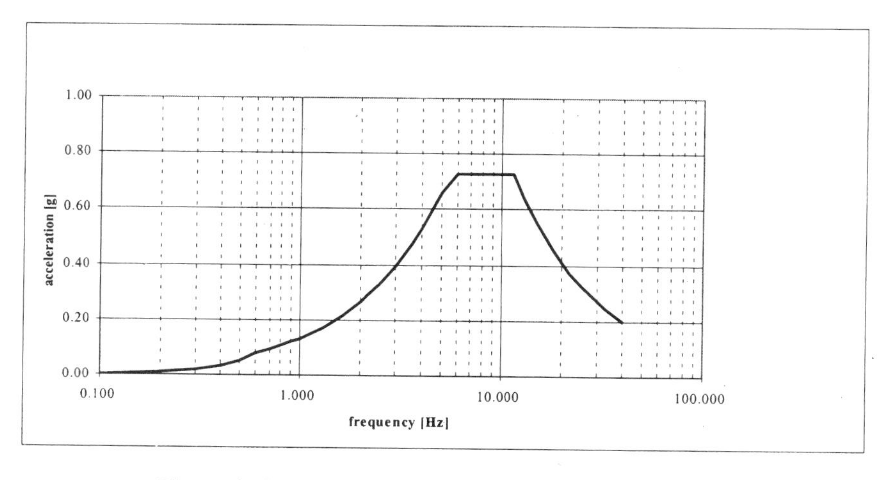

Figure 6 : 0.2g PML Secondary Response Spectrum

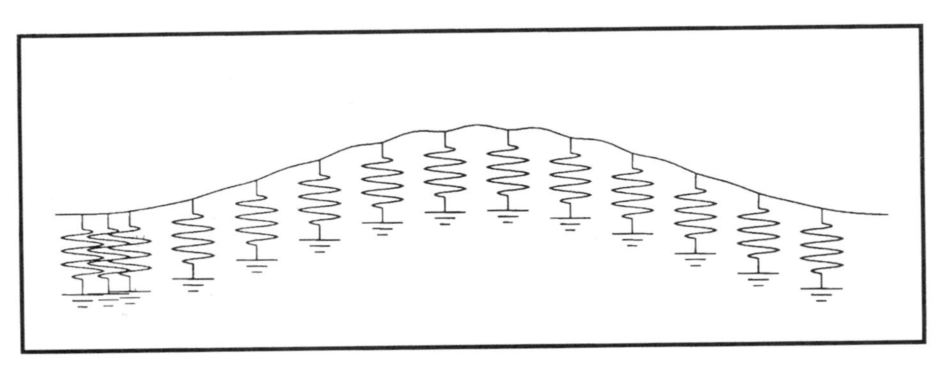

Figure 7a : 5.7Hz Mode shape of pipe behaving as a multi-span beam – flexible saddles

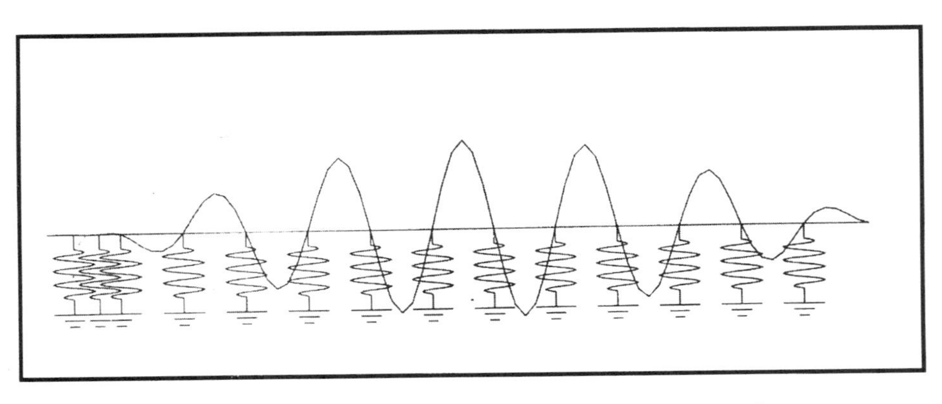

Figure 7b : 12 Hz Mode shape of pipe behaving as multi-span beam

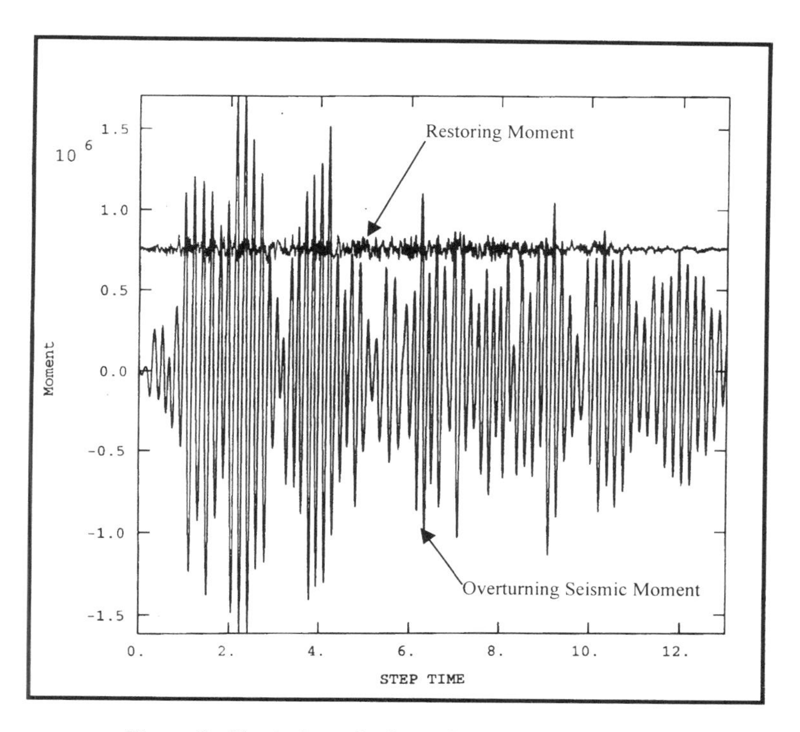

Figure 8 : Typical results from the ABAQUS analysis Showing saddle overturning moment v. restoring moment

C571/011/99

Applying structural reliability methods to ageing pipelines

A FRANCIS, A M EDWARDS, and **R J ESPINER**
BG Technology, Loughborough, UK
G SENIOR
Transco, Newcastle-upon-Tyne, UK

ABSTRACT

The use of Structural Reliability Analysis (SRA) to justify continued safe operation of pipelines is increasing. Indeed, the technique has recently been used by BG to demonstrate that 900 km of high pressure pipelines in the UK Transco National Transmission System could operate safely at a stress level of 78% SMYS, which is above the code requirements.

The overall aim of the structural reliability based method is to assess the fitness for purpose of a pipeline based on a calculated failure probability. It takes account of natural variations in pipeline dimensions and material properties and directly addresses the mechanisms that lead to failure.

This paper presents the basic principles of SRA and gives a detailed description of each of the key elements of the approach including relevant mathematical formulations where appropriate.

Particular focus is given to the application of the approach to ageing pipelines and some detailed guidance is given on the use of the technique to extend pipeline life. To this end the definition of pipeline life is given detailed consideration. Guidance is given on how the technique may be used to extend life by removing inherent conservatism, introducing and quantifying the effects of failure mitigating activities, or a combination of these.

NOMENCLATURE

a_g	Gouge depth
a_{corr}	Corrosion defect depth
a_0	Defect depth at commissioning
a_{seam}	Seam weld defect depth
a_x^{crit}	Critical depth of defect due to mechanism x
B	trawl board height
D	Dent depth

D^{crit} Critical dent depth
E Young's modulus
f Frequency of pressure cycles
f_i Damage frequency / defect density for failure mode i
F Externally applied force
F_{axial} Axially applied force
F^{crit} Critical externally applied force
F_{line} Trawl line breaking force
$G(.)$ General limit state function
h_s Free span height
H Heaviside step function
K_r Fracture parameter
K_{IC} Fracture toughness
l_g Gouge axial length
l_{corr} Corrosion defect axial length
l_{seam} Fatigue crack axial length
l_{tooth} Length of excavator tooth
L_A Acceptable life
L_E Expected life
L_N Nominal design life
$LN(.)$ Lognormal pdf
M Folias factor
$N(.)$ Normal pdf
p Net operating pressure
$p(.)$ Probability density function
P_f Cumulative probability of failure
Q Length correction factor
R External radius
R_p Puncture resistance
SMYS Specified minimum yield strength
S_r Collapse parameter
t time
U_{lift} Displacement to detach trawl gear
w Wall thickness
w_{nom} Nominal wall thickness
w_{tooth} Width of excavator tooth
x Dent depth limit
X Fatigue crack growth function
Y Stress intensity factor (SIF)
Y_b Normalised SIF for edge cracked strip in bending
Y_m Normalised SIF for edge cracked strip in tension
Δp Pressure range
ε Failure probability criterion
ε_{LB} Lower bound failure probability criterion
ϕ Central displacement function
σ_b Bending stress
σ_f Flow stress
σ_h Hoop stress
σ_m Membrane stress
σ_u Ultimate tensile strength
σ_y Yield strength
ω Distributed load per unit length

1 INTRODUCTION

Pipeline design codes were originally developed when relatively little experience of operating pipelines was available. Due to the paucity of information, the codes were based on a number of assumptions that would lead to conservative designs although the level of conservatism was largely unknown.

However, operational experience gained over the last 30 years or so has provided a wealth of information. It is now well known that high pressure gas pipelines designed to traditional codes have an impressive safety record, exceeding that of many other industries. Indeed, there have been no fatalities due to the failure of a high pressure gas pipeline in the UK during some 500,000 kilometre years of operation. There is also considerable information available from in-line inspection tools that describe the size and type of damage that may occur and the frequency of occurrence. Based on this significant increase in available information, there is now an increasing adoption of the belief that that some of the conservatism can be relaxed without compromising safety. The reduction of conservatism would bring considerable financial benefit to pipeline operators in the form of life extensions, pressure uprating and reduced inspection and maintenance, for instance. However, the acquisition of data of the form described above is not wholly sufficient to justify the relaxation of conservatism. The available data strongly suggests that current designs and operating scenarios are conservative but it does not provide any guidance on either the source or the magnitude of the conservatism. To this end recourse is made to structural reliability analysis (SRA) (1-12).

SRA is used to determine the likelihood of failure for the pipeline, as a function of time, based on a probabilistic analysis of the conditions that lead to failure.

This paper presents a detailed description of SRA and indicates how it might be applied to justify life extension without compromising safety. The application of the technique to justify safe pressure uprating has been described elsewhere (2-7, 11).

2 STRUCTURAL RELIABILITY ANALYSIS

SRA comprises the six basic elements, namely:

- Limit States
- Failure Modes
- Limit State Functions
- Data Analysis
- Evaluation of Failure Probability
- Assessment of Results

A brief description of each of these elements is given below.

2.1 Limit states

A limit state is defined as the state of a structure when it no longer satisfies a particular design requirement. For pipelines, limit states are defined in accordance with two specific design requirements:

1. The **Ultimate Limit State** (ULS) is the state at which the pipeline cannot contain the fluid it is carrying. This limit state may have safety or environmental implications. Examples of this limit state are leaks and ruptures. The Ultimate Limit State is generally associated with failure modes involving defects.

2. The **Serviceability Limit State** (SLS) is the state at which the pipeline no longer meets the full design requirements but is still able to contain the fluid, e.g. can no longer pass sufficient fluid and/or maintenance tools. This limit state has no direct safety implications. Examples of this limit state include permanent deformation due to yielding or denting. The Serviceability Limit State is generally associated with failure modes of defect free pipe.

2.2 Failure modes

A failure mode is the mechanism that causes the pipeline to reach a limit state. The failure modes for a particular pipeline depend on the operating environment and conditions, pipeline protection methods employed and the inspection and maintenance strategy adopted.

2.3 Limit state functions

A limit state function is a mathematical relationship between the relevant parameters characterising a particular failure mode that exists when the pipeline has reached its limit state. It is either a theoretical, semi-empirical, or empirical model that predicts the onset of failure of the pipeline. If there are n parameters, x_1, x_2,...x_n, the limit state function may be expressed in the form

$$G(x_1, x_2,...x_n) = 0 \qquad [1]$$

and failure will occur if the inequality

$$G(x) \leq 0 \qquad [2]$$

is satisfied.

2.4 Data analysis

In any practical engineering circumstances, each of the input values to a limit state model, and indeed the model itself, are subject to uncertainty. Uncertainties are accounted for in a structural reliability analysis by describing variables in statistical terms.

For each limit state function, the variability in the sensitive parameters has to be quantified by data analysis and ultimately the construction of probability density functions. This is achieved by performing appropriate statistical analyses of the data available from sources including construction records, test certificates and inspection records. The outputs of these calculations are mathematical functions describing the likelihood of occurrence of specified values of particular parameters.

2.5 Evaluation of failure probability

The probability density functions for each sensitive parameter are used in conjunction with the limit state functions to determine the probability of failure. When the stochastic parameters are independent the probability of failure, given that a defect exists, for each of the i credible failure modes is given by

$$P_{f_i}(t) | \, defect = \int_{x_1} p[x_1(t)] \int_{x_2} p[x_2(t)] \ldots\ldots \int_{x_n^{crit}} p(x_n(t)) \; dx_n \ldots dx_2 \; dx_1 \qquad [3a]$$

for time dependant failure modes within the interval (0, t) and

$$P_{f_i} | \, defect = \int_{x_1} p(x_1) \int_{x_2} p(x_2) \ldots\ldots \int_{x_n^{crit}} p(x_n) \; dx_n \ldots dx_2 \; dx_1 \qquad [3b]$$

for randomly occurring failure modes, where x_n^{crit} is the solution of equation [1] for given values of x_1, x_2, ..., x_{n-1}.

The probability of failure within the interval (0, t), per unit length, due to each of the i failure modes is given by

analysis and full scale testing. These guidelines, which are based on plastic collapse, are central to the derivation of the limit state function used by BG Technology.

The limit state function for corrosion defects relates the defect depth at failure to the defect length and hoop stress and is given by (11)

$$\frac{a_{corr}^{crit}}{w} = \frac{1 - \frac{\sigma_h}{\sigma_u}}{1 - \frac{\sigma_h}{\sigma_u} Q^{-1}} \quad [19]$$

where a_{corr}^{crit} is the depth of corrosion at failure and Q is a length correction factor given by

$$Q = \left[1 + 0.31 \left(\frac{l_{corr}}{\sqrt{2Rw}} \right)^2 \right]^{\frac{1}{2}} \quad [20]$$

where l_{corr} is the axial length of the corrosion defect. Hence, equation [19] may be used to determine the critical depth of a corrosion defect and the functional dependence

$$p = p\left(a_{corr}^{crit}, l_{corr}, R, w, \sigma_u\right) \quad [21]$$

implicitly exists.

3.3.5 Fatigue crack growth of construction defects

Defects and damage can be introduced into the parent plate and the seam and girth welds during manufacturing and construction. The most critical in terms of fatigue are crack-like welding defects introduced into the longitudinal seam weld.

The failure of a crack-like defect in a pipeline is predicted using fracture mechanics, however it is assumed that failure of construction defects is dominated by plastic collapse. The plastic collapse criterion developed by Battelle (20) for part wall defects in linepipe given by

$$\frac{a_{seam}^{crit}}{w} = \frac{1 - \frac{\sigma_h}{\sigma_f}}{1 - \frac{\sigma_h}{\sigma_f} M^{-1}} \quad [22]$$

is thus used to predict the critical defect depth, a_{seam}^{crit}. The Folias factor, M, is given by (15)

$$M = \left(1 + 0.26 \left[\frac{l_{seam}}{\sqrt{Rw}} \right]^2 \right)^{\frac{1}{2}} \quad [23]$$

where l_{seam} is the seam weld defect length.

It is assumed that the length, l_{seam}, of a given crack does not increase but that the crack depth, a_{seam}, will increase with time under the influence of a cyclic hoop stress. The fatigue crack growth is defined by a function X, based on the well known Paris fatigue crack growth law (21), such that the crack depth, $a_{seam}(t)$, is given by

$$a_{seam}^{crit}(t) = X\left[a_0, \Delta p, f, t, Y(a, w)\right] \qquad [24]$$

where Δp and f are the range and frequency of pressure cycling respectively, a_0 is the depth of the defect at the time of commissioning (time zero). The stress intensity factor, Y(a, w), is found using the Newman-Raju solution (22) for an internal, axially orientated, semi-elliptical crack in a cylindrical pressure vessel.

Equation [24] may be used to determine the critical depth of a seam weld defect and the functional dependence

$$p = p(a_{seam}^{crit}, l_{seam}, \Delta p, R, w, \sigma_f, a_0, f, t) \qquad [25]$$

implicitly exists.

3.3.6 Buckling or collapse

Buckling or collapse of the pipeline due to trawl gear hooking will occur if

- the critical applied force, F^{crit}, (determined from the axial force and bending moment capacities of the pipeline) is less than the force required to break the line connecting the trawl gear to the trawler, F_{line}; and
- the critical central displacement of the free spanning pipe section is less than the displacement, U_{lift}, at which the trawl gear will become detached from the pipeline

The limit state function, $G(F^{crit})$ is thus expressed in the form

$$G(F^{crit}) = \phi(F^{crit}) - Min\{U_{lift}, \phi(F_{line})\} \qquad [26]$$

where the function $\phi(F)$ relates the applied load, F, to the central displacement, U, of the free spanning pipeline and is determined from approximations found in (23) and is given by

$$\phi(F) = \frac{F^2}{8\omega F_{axial}} - \frac{F}{2F_{axial}K}\tanh\left(\frac{FK}{4\omega}\right) \qquad [27]$$

and

$$U = \phi(F) \qquad [28]$$

where ω represents the distributed load per unit length on the pipeline, F_{axial} the applied axial force and K is given by

$$K = \sqrt{\frac{F}{EI}} \qquad [29]$$

where E is the Young's modulus of the pipeline steel and I is the second moment of area of the pipe cross section. The displacement U_{lift} is given by (24)

$$U_{lift} = 0.7B - 0.6R + 0.3(B + 2R)H(h_s - 0.7B) \qquad [30]$$

where B is the trawl board height, h_s is the free span height and H is the Heaviside step function. Failure will occur due to buckling or collapse when

$$G(F^{crit}) \leq 0. \qquad [31]$$

Hence, equation [26] may be used to determine the critical load due to trawl gear hooking that will lead to buckling or collapse of the pipeline and the functional dependence

$$F_{axial} = F_{axial}(F^{crit}, R, w, E, h_s, B, F_{line}) \qquad [32]$$

implicitly exists.

3.4 Data analysis

The physical parameters identified in the functional relationships for each limit state function are subject to natural variation and in principle should be treated as stochastic variables.

However, in practice only the physical parameters that the limit state function is most sensitive to and are subject to significant variation need to be treated as stochastic variables, the remaining parameters may be represented by nominal or conservative deterministic values.

For in-service pipelines there are numerous sources of information from which probability density functions to represent each of the sensitive parameters may be constructed. These include survey and inspection results, operating history, mill inspection records, hydrostatic test records, fluid composition and corrosion monitoring data. The relevance of each of these is discussed below.

3.4.1 Geometry and material parameters

The material and geometric parameters that are subject to significant variation between (and also within) linepipe sections are w, σ_y and hence σ_f, σ_u and K_{IC}. These parameters are therefore represented by probability density functions (pdfs) constructed directly or indirectly from data obtained from mill inspection records.

A large data set is required in order to construct a pdf that represents the parameter to a high level of confidence. From experience (25) however, it has been noted that these w and σ_y fit similar distributions and may be represented by the following generic distributions,

$$w \sim N(w_{nom}, 2.5\, mm) \qquad [33]$$

$$\sigma_y \sim LN(1.1SMYS, 0.044SMYS) \qquad [34]$$

where w_{nom} is the nominal wall thickness and SMYS is the specified minimum yield strength.

Significance tests should be performed on the generic distributions if limited pipeline specific data is available, this will enhance the confidence held in the pdfs. Also, the distribution may be further refined from consideration of the hydrostatic test pressure. This allows combinations of w and σ_y that would have failed at the hydrostatic test to be eliminated (8).

For conservatism, K_{IC} may be represented by the nominal minimum value corresponding to the nominal minimum Charpy value given by the appropriate specification.

3.4.2 Environmental parameters

For relevant offshore applications the trawl board height, B, may be characterised from knowledge of the trawl gear type used in the vicinity of the pipeline. Example data is presented in (24).

A distribution for h_s may be constructed from data obtained from a sonar survey along the length of the pipeline. Previous studies have indicated that h_s is described by a Lognormal distribution. This information should be incorporated with knowledge of the seabed conditions. Information on likely seabed movement may be obtained from consideration of successive sonar surveys of the pipeline under study or of other pipelines in the area.

3.4.3 Damage Parameters

Experience (2-11) has shown that the probability of failure is more sensitive to the distributions assumed for damage parameters than for the geometry and material properties. The remaining life of a pipeline is most sensitive to those failure modes involving degradation mechanisms, i.e. those that have a time dependent failure probability.

Probability density functions to represent damage parameters, e.g. a_g, l_g, D, a_{corr} and l_{corr}, may be constructed from consideration of in-line inspection records and defects reported through other means. Previous studies have indicated that a_g, D and a_{corr} are represented by Weibull distributions whilst l_g and l_{corr} may be represented by offset logistic distributions (12).

Data to construct the rate of growth of a_{corr} and l_{corr} is obtained from consideration of successive in-line inspections or from corrosion monitoring equipment including weight loss coupons.

It is recognised that not all pipelines have been subjected to an in-line inspection or have active corrosion monitoring. Suitable data may however be obtained from pipeline systems operated under similar conditions to that under study.

3.5 Probabilistic calculations

For a life extension study, equations [4a] and [4b] are evaluated over a range of time intervals for each credible failure mode. The total probability of failure is given by equation [5]. A typical profile of the total probability of failure as a function of time is given in Figure (1).

3.6 Interpretation of results

The objective of the methodology described above is to evaluate the probability that the pipeline will fail within a given time period. The simplest analysis will give the probability of failure within the time period between the date of commissioning and the end of the design life. A slightly more complex analysis will give the probability of failure between the current time and the end of the design life, taking into account previous operating history including the effects of inspection and maintenance. Other analyses can be conducted between the above start times and some arbitrary time in the future. Whichever, type of analysis is to be conducted a decision has to be made on the acceptability of the pipeline based on the results.

There are currently no formal criteria on which to determine the safe design life, although work is in progress under a Joint Industry Project (13) towards this end. However, the SRA approach has been used successfully to justify increasing the maximum allowable operating pressure (MAOP) of pipelines.

The remainder of this section provides an insight into how the results might be interpreted and suggests how life extension may be achieved based on these interpretations.

3.6.1 Definition of Pipeline Life

Before going any further it is useful to ask the intuitively obvious question: What is meant by the term 'design life'? Unfortunately, the answer to this question is not obvious. Definitions such as 'The time period from installation until permanent decommissioning of the pipeline system' (26) are not particularly helpful since the questions: 'how long should this time period be?' and 'how is it determined?' still remain unanswered.

Pipeline design codes often stipulate the requirements to achieve a fatigue life of 40 years, say. This is done by placing limits on the allowable number of stress cycles of a given range within the 40 year period. Offshore pipeline design codes include guidance on the inclusion of a corrosion allowance, which has implications on life. However, it is not possible to guarantee a life based on this allowance, the allowance is used to account for the uncertainty in corrosion growth that may take place over a notional life. Also, despite design codes stipulating external coating and cathodic protection requirements, external corrosion is also known to affect pipelines. There is thus need to exercise caution when treating fatigue life and design life as synonymous. In view of this a detailed consideration of pipeline life is given below.

3.6.1.1 Expected life

The expected life, L_E, can be precisely defined as

$$L_E = \int_0^{\infty} [1 - P_f(t)] dt \qquad [35]$$

where $P_f(t)$ is the cumulative probability of failing in the time interval (0, t) given by equation [5]. This quantity is used extensively in other fields of application. Not least of these is the evaluation of human life expectancy, which is one of the key parameters used by life insurance companies to determine premiums. Insurance companies effectively guarantee to the shareholders, the remaining lives of those insured, based on life expectancy that is often estimated by taking into account factors such as gender, eating, drinking and smoking habits. The numbers of lives insured often run into thousands and tens of thousands which means that, although in cases of premature death the company will make a loss, overall they will make a profit.

3.6.1.2 Acceptable Life

Pipeline design codes are effectively required to guarantee the lives of pipelines. However, in this case, the life expectancy is of little value since relatively few pipelines are operated and the occurrence of a single failure can have severe socio-economic consequences. For this reason the 'acceptable life' of the pipeline has to be selected such that a failure is very unlikely within that acceptable life. The implication of this is that the expected life is likely to be an order of magnitude greater than acceptable life.

A useful definition of acceptable life, L_A, would thus be

$$P_f(L_A) \leq \varepsilon \qquad [36]$$

where ε is suitably small. There is currently no formally agreed value of ε. It is not the intention of this paper to prescribe such a value but rather to suggest a means of doing this. However, it is worth exploring various other descriptions of life before considering the implications of the above inequality in detail.

3.6.1.3 Nominal Design Life

The nominal design life, L_N, is essentially the life which implicitly satisfies the equation [36]. However, it is not determined by explicit consideration of failure probability but rather by prescribing design and operational requirements, based on experience and engineering judgement, which have resulted in very few or no failures. However, it is generally recognised that the various prescribed requirements such as those on the number and range of stress cycles and the corrosion allowance on wall thickness, are conservative and that the nominal design life is thus a lower bound estimate of the acceptable life. It follows that a lower bound estimate of acceptable failure probability, ε_{LB}, can be obtained from

$$P_f(L_N) = \varepsilon_{LB} \qquad [37]$$

However, it is recognised that operation to design codes leads to an inconsistent consideration of safety (4), therefore, the right-hand side of equation [37] may be different for two nominally identical pipelines.

Also, caution must be exercised when evaluating the left-hand side of the above equation. Failure mitigating activities such as aerial surveillance, inspection and maintenance are undertaken throughout the life of the pipeline. These have the effect of reducing the failure probability and should therefore be taken into account during the evaluation of $p_f(L_N)$. It is also the consideration of these effects that leads us to the concept of economic life.

3.6.1.4 Economic Life

It is possible to reduce the failure probability within a given time interval, thus extending the 'life', by introducing more frequent and more stringent failure mitigating activities. However, there is an associated cost in doing this. The economic life of the pipeline can therefore be defined as the life which is achievable by the introduction of further mitigating activities but for which the cost of a further increase would not be economically viable. The economic viability is dependent on the revenue that may be realised from life extension; this depends on the demand for the transported product, the amount of the product available, and the cost effectiveness of the measures that may be taken to manage the probability of failure. It should be noted, however, that even in situations where the supply and demand are increasing the economic life does not become indefinite; this is because the situation will eventually be reached where the cost of failure mitigation would exceed that of installing a new pipeline or abandoning the field.

3.6.2 Life extension

The discussion given above has illustrated that the acceptable life has to be much shorter than the expected life in order to achieve an acceptable level of safety.

Safe operation has traditionally been achieved by operating to a nominal life. The nominal life is generally based on a conservative approach to safety and will usually represent a lower bound estimate of the acceptable life; however, this statement requires justification if safe operation is to be extended beyond nominal life.

The acceptable life can be extended by introducing mitigating activities at cost. However, a time may be reached when such cost is greater than the benefits and this determines the economic life. The determination of economic life requires the cost of mitigation and expected revenue to be taken into account.

The implication on cost of operating a pipeline beyond the nominal life is illustrated in Figure 2. It is seen that life extension essentially falls into two categories:

(1) The extension of pipeline operation beyond the nominal life based on maintaining acceptable levels of safety without cost implication.

(2) The extension of acceptable life taking into account all costs and thus determining the economic life.

Both of the above can be addressed using the SRA technique. However, a detailed analysis of costs including supply and demand forecasts is required for the latter; this requires much pipeline specific data and is therefore not addressed here. The approach adopted is to quantify the maintenance and inspection strategy required in order to extend the acceptable life beyond the nominal design life.

There are essentially two elements to the approach. The first of these is to determine an acceptable probability of failure, i.e. to define ε. This can be done by evaluating the probability of failure between the time of commissioning and a range of times up to the current time. Such evaluations should include the effects of any inspection and maintenance. This information is clearly more meaningful, and of more value, for pipelines which have operated for a significant proportion (>75%, say) of their nominal design life. In situations where this is not true it may be more useful to evaluate the probability of failure of similar pipelines which have operated for a significant proportion of the nominal design life or to evaluate the probability of failure between the time of commissioning and the nominal design life as discussed above.

The second element is to evaluate the probability of failure within the time interval between the current time and desired acceptable lives taking into account all previous operating history, if any. A comparison between these calculated failure probabilities and those associated with the life to date, which may be implicitly regarded as acceptable, may be used as a justification for extension of life beyond the nominal design life. Where future failure probabilities are considered to be too large, the effects of further mitigation should be evaluated and appropriate recommendations made. A consideration of such mitigation is given below.

4 OPTIMISATION OF FAILURE MITIGATING ACTIVITIES

Inspection and maintenance activities are conducted in order to achieve an acceptable or desired level of reliability (and hence safety). Inspection activities reduce the probability of failure (and hence increase reliability) through the reduction of uncertainty in the quantity and size distribution of defects at the time of the inspection. Maintenance activities reduce the probability of failure through the removal of defects that may grow to a critical size within an unacceptable time interval.

The optimum maintenance/inspection strategy should address the most likely failure mode, i.e. the mode that provides the greatest contribution to the failure probability of the pipeline within the interval (0, t), evaluated from [4a] and [4b].

A discussion on how structural reliability analysis may be used to quantify the reduction in failure probability due to performing the more common maintenance and inspection activities is given below.

4.1 In-line inspection (MFL pig)

The Magnetic Flux Leakage (MFL) pig detects metal loss defects including gouges and corrosion and dents. If a defect greater than the adopted repair criterion is reported it will be repaired.

However, external interference defects (i.e. dent/gouges and dents) are introduced randomly into a pipeline during service and are assumed not to be subject to further degradation. Therefore if the pressure remains approximately constant, a defect will either fail when it is introduced or remain safely in the pipeline. The in-line inspection will not therefore significantly reduce the probability of failure due to external interference during normal service.

Benefit from the in-line inspection is derived from the removal of uncertainty in the quantity and size distribution of corrosion metal loss defects. This has the effect of reducing the calculated probability of failure due to internal and external corrosion.

Inspection using the MFL pig should therefore be considered if the most credible failure mode is internal or external corrosion and the level of failure probability accepted by the operating company, given by [36], will govern the scheduling of the next inspection. However, if the most significant failure mode is external interference then it is evident that a more optimum solution to reducing failure probability and thus extending pipeline life may exist.

4.2 Aerial surveillance

The objective of aerial surveillance is to detect excavating activities within the vicinity of the pipeline and then to take appropriate action, if necessary, to prevent impact. The action taken is dependent on the proximity and type of activity. In the UK, helicopter surveillance is used and in situations of most concern the helicopter is landed and the activity is halted immediately. Incidences that may lead to a puncture, critical dent/gouge defect or critical dent may therefore be prevented by aerial surveillance and thus the probability of failure due to these modes is reduced if aerial surveillance is conducted along the pipeline length.

Onshore pipelines operated according to IGE/TD/1 (27) are surveyed every 14 days. However, the majority of activities occurring near a pipeline last for a period much shorter than 14 days. Therefore, there is the potential for significant activity close to a pipeline to go undetected.

It is clear that the reduction in the failure probability is proportional to the frequency of aerial inspection. Increasing the surveillance frequency however leads to a continuous increase in expenditure and more cost-effective solutions to reducing the probability of failure and thus extending the pipeline life may exist if external interference is the most credible failure mode. Examples include participation in a one-call system, on-site supervision of contractors, satellite surveillance and pipeline marking. The effect of each of the mitigating measures may be accounted for in the structural reliability analysis by modifying the parameter f_i in equations [4a] and [4b] by the applicable factor.

4.3 Side scan sonar survey

A side scan sonar survey is conducted offshore in order to generate an image of the seafloor conditions and will therefore detect uncovered pipe sections, free-spanning pipe sections and any objects within close proximity to the pipeline.

Thus the survey may reduce the probability of failure due to trawl gear hooking, trawl board impact and vortex induced vibrations through the removal of uncertainty in the quantity and size distribution of uncovered and free spanning sections. However, in environments where trawling activity and the current velocity is low the impact of the survey on the increasing he pipeline life may be negligible.

The level of failure probability accepted by the operating company, given by equation [36], will govern the scheduling of the next inspection.

5 CONCLUSIONS

The application of structural reliability analysis to pipelines has been describedl. The relevant mathematical formulations appropriate to both onshore and offshore pipelines have been given where appropriate. A particular focus has been given to the application of SRA to extending pipeline life and to this end various definitions of life have been given. These are expected life, nominal life, acceptable life and economical life. The implications of each of these definitions has been discussed and guidance has been given on how SRA be used to extend life by either reducing inherent conservatism, introducing and quantifying the effect of further mitigation or a combination of these.

6 REFERENCES

1. Ditlevsen, O. & Madsen, H., *"Structural Reliability Methods"*, J. Wiley, 1996.
2. Francis, A., Batte, A.D. & Haswell, J.V., "*Probabilistic Analysis to Assess the Safety and Integrity of Uprated High Pressure Gas Transmission Pipelines*", Institution of Gas Engineers Annual Conference, Birmingham, UK, April 1997
3. Francis, A., Espiner, R.J., Edwards, A.M., Cosham, A. & Lamb,M., "*Uprating an In-service Pipeline Using Reliability-based Limit State Methods*", 2nd International Conference on Risk Based & Limit State Design & Operation of Pipelines, Aberdeen, UK, May 1997
4. Francis, A. & Senior, G., "*The Applicability of a Reliability-based Methodology to the Uprating of High Pressure Pipelines*", Institution of Gas Engineers Midlands Section Meeting, Hinckley, UK, March 1998
5. Francis, A., Espiner, R.J., Edwards, A.M., & Senior, G., "*The Use of Reliability-based Limit State Methods in Uprating High Pressure Pipelines*", International Pipeline Conference, Calgary, Canada, June 1998
6. Senior, G., Francis, A. & Hopkins, P., "*Uprating the Design Pressure of In-service Pipelines Using Limit State Design and Quantitative Risk Analysis*", 2nd International Pipelines Conference, Istanbul, Turkey, December 1998
7. Francis, A., Edwards, A.M. & Espiner, R.J., "*Reliability Based Approach to the Operation of Gas Transmission Pipelines at Design Factors Greater Than 0.72*", 17th International Conference on Offshore Mechanics and Arctic Engineering, Lisbon, Portugal, July 1998
8. Espiner, R.J. & Edwards, A.M., "*An Investigation of the Effectiveness of Hydrostatic Testing in Improving Pipeline Reliability*", 3rd International Conference on Risk Based & Limit State Design & Operation of Pipelines, Aberdeen, UK, October 1998
9. Lamb, M., Francis, A. & Hopkins, P., "*How do you Assess the Results of a Limit State Based Pipeline Design*", 3rd International Conference on Risk Based & Limit State Design & Operation of Pipelines, Aberdeen, UK, October 1998
10. Batte, A.D., Francis, A. & Fu, B., "*Extending the Operational Performance of Pipelines*", International Gas Research Conference, San Diego, USA, November 1998

11. Espiner, R.J., *"Uprating of In-Service Transmission Pipelines using Structural Reliability Based Methods"*, Institute of Gas Engineers North of England Section", UK, January 1999.

12. Espiner, R.J., Edwards, A.M. & Francis, A., *"Structural Reliability Based Approach to Uprating a Sub-Sea High Pressure Gas Pipeline"*, 18th International Conference on Offshore Mechanics and Arctic Engineering, Newfoundland, St. Johns, July 1999.

13. BG Technology, Andrew Palmer & Associates, CFER Technologies Inc, JP Kenny Ltd, Advanced Mechanics & Engineering Ltd., *"Proposal – Joint Industry Project to Develop Guidance for Limit State, Reliability and Risk Based Design and Assessment of Onshore Pipelines"*, January 1999.

14. British Standards Institution, *"Guidance on Methods for Assessing the Acceptability of Flaws in Fusion Welded Structures"*; PD6493, 1991.

15. Folias, E.S., *"Fracture of Nuclear Reactor Tubes"*, Paper C4/5, SmiRT III, London, 1975

16. Driver, R.G. & Zimmerman, T.J.E., *"A Limit States Approach to the Design of Pipelines for Mechanical Damage"*, 17th International Conference on Offshore Mechanics and Arctic Engineering, Lisbon, Portugal, July 1998.

17. Corbin, P. & Vogt, G., *"Future trends in Pipelines"*, Proceedings Banff 1997 Pipeline Workshop: Managing Pipeline Integrity – Planning for the Future, Banff, Alberta, Canada, 1997

18. Corder, I. & Chatain, P., *"EPRG Recommendations for the Assessment of the Resistance of Pipelines to External Damage"*, EPRG/PRC, 10th Biennial Joint Technical Meeting on Line Pipe Research, Cambridge, UK, April 1995.

19. Batte, A.D., Fu, B., Kirkwood, M.G. & Vu, D., *"New Methods for Determining the Remaining Strength of Corroded Pipelines"*, 16th International Conference on Offshore Mechanics and Arctic Engineering, Yokohama, Japan, April 1997

20. Kiefner, J.F., *"Fracture Initiation"*, Paper G, 4th Symposium on Linepipe Research, AGA, Dallas, Texas, 1969

21. Paris, P.C. & Erdogan, F.A., *"A Critical Analysis of Crack Propagation Laws"*, Journal of Basic Engineering Transactions of ASME, Vol 85D, No. 4., 1963.

22. Newman, J.C. & Raju, I.S., *"Stress Intensity Factor Equations for Internal Surface Cracks in Cylindrical Pressure Vessels"*, Transactions of ASME, Journal of Pressure Vessel Technology, Volume 102, November 1980.

23. Young, W.C., *"Roark's Formulas for Stress & Strain"*, 6th Edition, McGraw-Hill International Editions, 1989.

24. Fyrileiv, O., Spiten, J., Mellem, T. & Verley, R., *"DNV '96, Acceptance Criteria for Interaction Between Trawl Gear And Pipelines"*, 16th International Conference on Offshore Mechanics and Arctic Engineering, Yokohoma, Japan, April 1997.

25. Jaio, G., Sotberg, T., Bruschi, R. & Igland, R.T., *"The SUPERB Project: Linepipe Statistical Properties and Implications in Design of Offshore Pipelines"*, 16th International Conference on Offshore Mechanics and Arctic Engineering, Yokohama, Japan, April 1997.

26. Anon., *"Rules for Submarine Pipeline Systems"*, Det Norske Veritas, 1996.

27. Anon, *"Steel Pipelines for High Pressure Gas Transmission"*, IGE/TD/1 Edition 3, 1993.

28. Corder, I., *"The Application of Risk Techniques to the Design and Operation of Pipelines"*, C502.016, IMechE 1995.

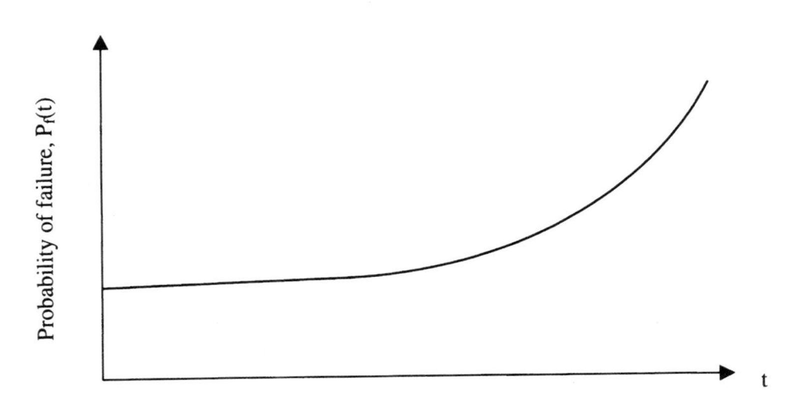

Figure 1: Typical profile for total failure probability as a function of time

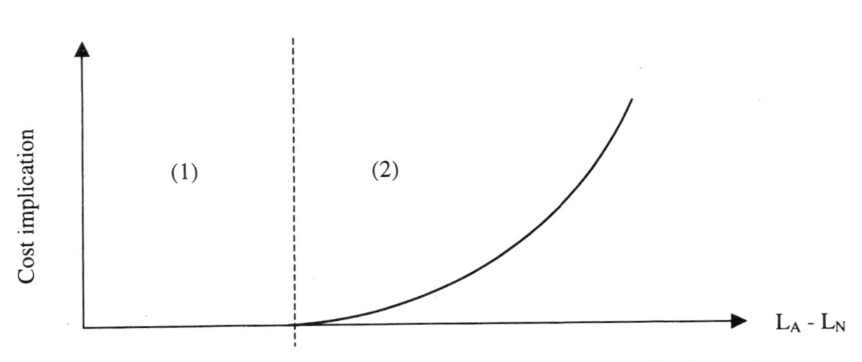

Figure 2: Schematic of the cost implications of pipeline operation beyond the nominal design life indicating two categories of life extension

C571/041/99

A strategy for the repair and rehabilitation of onshore and offshore pipelines

P HOPKINS and **D BRUTON**
Andrew Palmer and Associates, Newcastle-upon-Tyne, UK

ABSTRACT

Major transmission pipelines in the world are approaching the end of their design lives. They will require repair and rehabilitation of damage and defects if they are to remain in service.

Pipeline operators will usually inspect their pipeline prior to any rehabilitation programme, and assess the significance of defects detected in the pipeline. This assessment may be followed by repairs to, or rehabilitation of, the pipeline.

This paper presents a strategy for linking all the stages of pipeline inspection, assessment, and repair and rehabilitation, which is applicable to onshore and offshore pipelines.

1. INTRODUCTION - OVERVIEW OF REPAIR AND REHABILITATION

There are millions of kilometres of transmission pipelines around the world; the oil and gas transmission system in Western Europe alone is over 150,000 km in length. These pipelines are very safe structures (1-3), but occasionally an operator will detect, or become aware of, defects in his or her pipeline. In the past, this has usually led to expensive shut-downs and repairs, but in recent years there has been an increasing use of fitness-for-purpose methods to assess these defects, and cost effective repair and rehabilitation methods.

However, the inspections, fitness-for-purpose methods, and rehabilitation methods, are often treated in 'isolation', e.g. a defect is assessed and a recommendation made, without thought of repair strategy, long term inspection costs, impact on overall reliability, etc..

Therefore, pipeline operators are now looking for companies to not only inspect for, or assess the significance of, a defect, but to provide a 'turnkey' service, that includes inspection, assessment, repair and rehabilitation.

Accordingly, a robust strategy for this service is needed by the industry. This paper presents such a strategy, which produces the plan outlined in Figure 1.

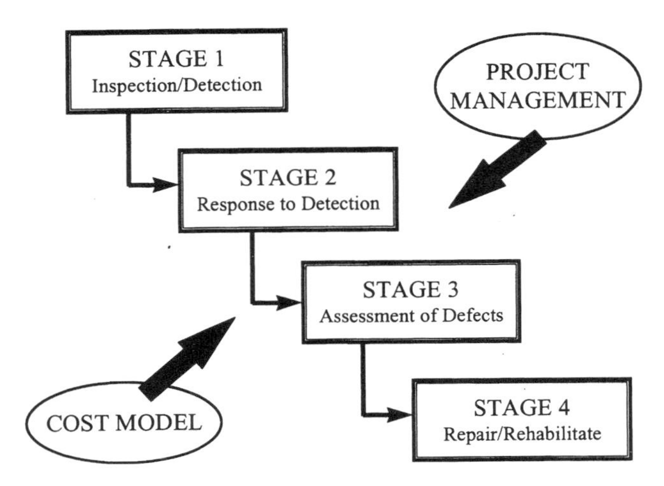

Figure 1. The four stages of a rehabilitation and repair plan.

1.1 Repair and rehabilitation - The Need

There are 50 years of proven oil & gas supplies in the world. However, many oil & gas fields' infrastructure are at the end of their design life but they still have 25 or even 50 years of production left. Examples of ageing infrastructures are:

i. 50% of Russia's pipeline infrastructure will be at the end of its design life in 15 years time (4),
ii. 40% of the Brazilian pipeline system is at the end of its design life,
iii. over 50% of the 1,000,000 km USA oil and gas pipeline system is over 40 years old. Many 1,000s of km of lines are being replaced, repaired or rehabilitated in the USA every year; the US Department of Transportation estimates that 80,000 km of pipelines will require rehabilitation in the next 10 years (5).

1.2 Repair and rehabilitation - Drivers

Repair or rehabilitation projects are not solely driven by the choice of repair or rehabilitation method. They are driven by economics, urgency or engineering considerations:

a. COST - for example, the cost of the repair clamp for a damaged offshore pipeline is negligible compared to the cost of the vessel that has to be hired to install the repair.
b. URGENCY - a catastrophic failure (effecting people or environment) has such a devastating effect on public relations, etc., that the least of a company's worries is the

type of repair/rehabilitation. Cost to public image, lost revenue and clean up costs are key considerations.

c. ENGINEERING - the 'engineering' associated with any work on a pipeline may be the crucial consideration. For example, an offshore line that has lost its weight coating, and is floating, or an onshore line that is to be lifted out of a trench and recoated live, will require extensive engineering work.

1.3 Repair & rehabilitation - Approach

Pipeline rehabilitation should be part of the whole life costing of a pipeline, and it should be treated as a pipeline engineering/construction/overhaul (i.e. engineering, retrieval/trenching, section rehabilitation, backfill, inspection, etc.), requiring a wide engineering approach. This approach will always ensure a cost beneficial solution. For example, knowledge of new construction prices can show that rehabilitation is increasingly attractive as the pipeline diameter increases.

Table 1. Costs for rehabilitating pipelines compared to new builds (landlines) (6).

Pipe size	% of new construction
6-14-inch	70-130%
16-24-inch	50-90%
28-48-inch	25-50%

Smaller diameter pipe is cheap to buy, hence material costs of a new line are low. Certainly, pipeline diameters below 16 inch may be cheaper to replace rather than rehabilitate (7). Therefore, rehabilitation becomes attractive as the diameter increases, but this is a general statement, and will not apply in every case (see Figure 2). In this case we can see that a strategy incorporating regular inspection is the best financial solution. Clearly a cost model must be used to determine the best rehabilitation solution (see Section 1.6).

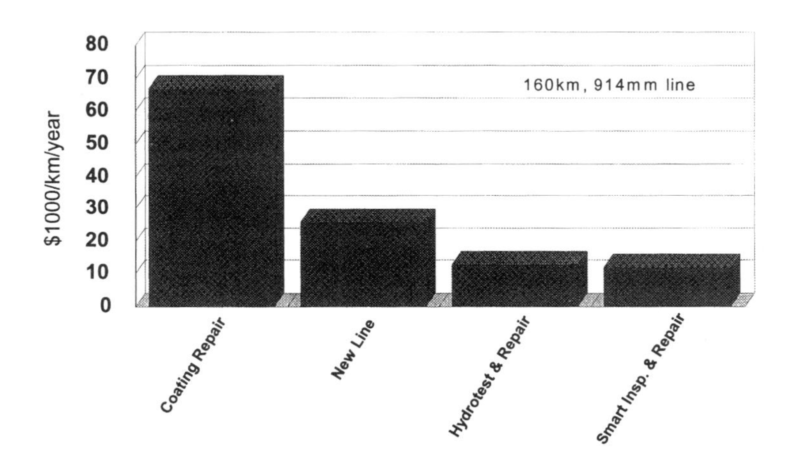

Figure 2. Differing costs for differing rehabilitation solutions (landlines) (8).

1.4 Repair & rehabilitation - Strategy

An operator will often be faced with an ageing pipeline, or one that is approaching the end of its life. He/she will need a strategy to ensure continued safe and efficient operation. The items in the strategy will be:

i. define goals (see following section),
ii. set up project team & collate data,
iii. build project cost model, to include whole life considerations (see Section 1.6),
iv. plan and implement inspection programme,
v. assess inspection results, and use cost model for optimum solutions,
vi. make line safe and repair/rehabilitate as necessary,
vii. put line back into service with future inspection and rehabilitation plan in place.

Figure 3 shows how a complete repair and rehabilitation plan can be built up. It includes

the fitness for purpose assessments, environmental response and hazards, etc., but also emphasises the importance of work before mobilising the repair teams.

It is this 'complete picture' that is essential, otherwise what looks like a relatively simple repair and rehabilitation programme can turn out to be a major loss making exercise.

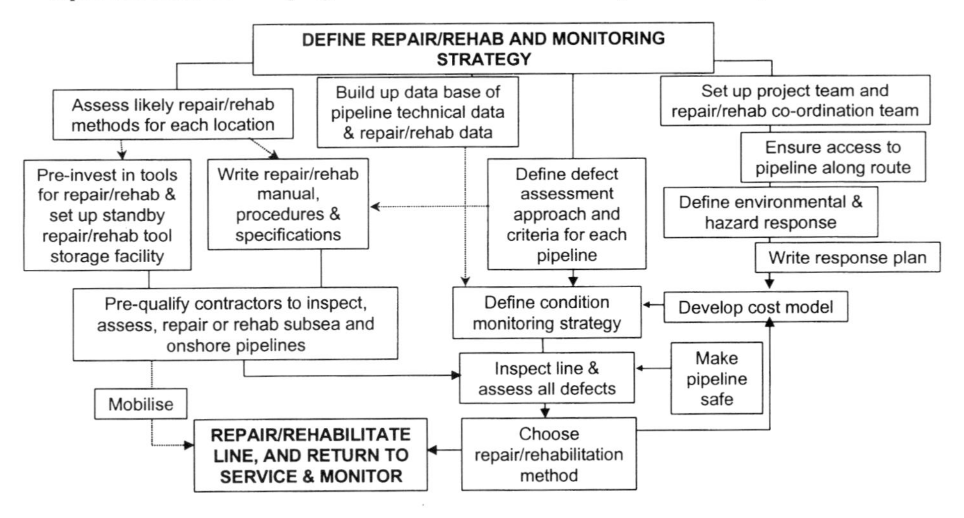

Figure 3. A complete repair and rehabilitation strategy (10)

1.5. Repair and rehabilitation - Project Management (9)

The project management principles used on new pipelines should be applied to pipeline rehabilitation:

i. project concept (goal statement),

ii. project study and definition (plan of execution),
iii. pre-construction (award contract),
iv. construction (turnover),
v. ongoing operations (inspection & maintenance).

Project management will co-ordinate many different services, including inspections, protection survey, coatings selection, etc.. Problems may include:

i. the scope of project is difficult to define as many uncertainties exist (e.g. location of corrosion),
ii. dealing with older pipelines, whose design, drawings, etc., may be difficult to locate/prove,
iii. projects are often 'fast-tracked' due to operational pressures,
iv. often the rehabilitation is on a live pipeline, with additional safety risks,
v. take up and removal of existing coating may be disposal problem,
vi. etc.!

1.6 Cost models

Any repair and rehabilitation programme needs to be costed. Figure 2 has shown how various options will incur widely varying costs. Figure 4 shows a simple cost model, based on pipeline failure probability. These models are a key element of the overall strategy.

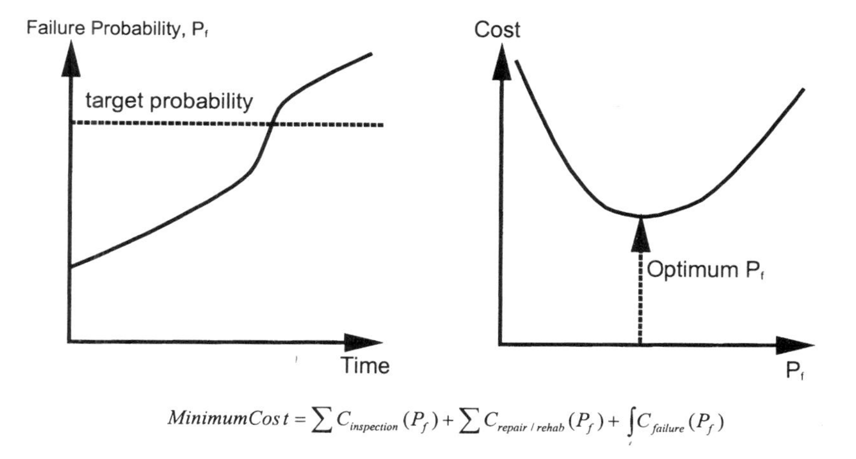

$$MinimumCost = \sum C_{inspection}(P_f) + \sum C_{repair/rehab}(P_f) + \int C_{failure}(P_f)$$

Figure 4. Example of cost model for repair and rehabilitation.

Figure 4 shows an operator setting a target failure probability, and ensuring that a rehabilitation programme does not take him/her past this target level. The total cost of rehabilitation in this model is taken as the sum of inspection and repair/rehabilitation costs (discrete values), and the cost of failure (variable). Iterations of the model will obtain the optimum cost solution. These type of cost models are an essential element of the strategy, Figure 3.

2. STRATEGY STAGE 1 - DETECTING DEFECTS IN PIPELINES

The detection of defects is the first stage of a repair and rehabilitation programme (Figure 1). Table 2 summarises pipeline inspection and maintenance methods, and shows which defects/damage these methods will detect.

These methods are dealt with in the literature, and therefore will not be dealt with here. However, it is important to emphasise that the quality of the methods used, and the reliability and accuracy of the data collated, will have a major effect on all the subsequent stages of a repair and rehabilitation programme.

Table 2. Pipeline inspection and monitoring methods

DEFECT/DAMAGE	METHOD[1]						
	AERIAL/ GROUND PATROLS /ROVs	*SMART PIGS*	*PRODUCT QUALITY*	*LEAK SURVEYS*	*GEOTECH SURVEYS & S. GAUGES*	*CP & COATING SURVEYS*	*HYDRO-TEST*
3rd Party Damage	P	R					R
Ext. Corrosion		R				P	R
Int. Corrosion		R	P				R
Fatigue/Cracks		R					R
Coatings	R[2]					P	
Materials/Construct Defects		R					R
Ground Movement	R				R		
Leakage	R	P		R			R
Sabotage/Pilfering	P						

3. STRATEGY STAGE 2 - RESPONSE TO DISCOVERING DEFECTS

Before attempting to assess the significance of any detected or suspected defect, it is first necessary to ensure that the pipeline is safe, and a plan exists to solve the whole 'problem', and not purely assess the detected defect. This action plan is in three parts:

1. **Reduce Pressure** - it may be necessary to reduce pressure to a safe level and plan pressure reductions for repair/rehabilitation,
2. **Critical Information** - Establish critical information for the engineering assessment (Section 4),

---

[1] Visual examinations, pressure monitoring and 'awareness' methods are not included.
P - 'proactive method, i.e. it prevents the defect/damage occurring. R - 'reactive' method, i.e. it detects the damage/defects after it has occurred.

[2] For subsea lines.

3. **Location Assessment** - Assessment of damage location to define access requirements and available repair and rehabilitation methods.

3.1 Reduce pressure

The pressure in the pipeline must be reduced to a safe level before repair. If the pipe is severed, or the defect is leaking, then it is essential to make the pipe and surroundings safe. There is often a need to apply immediate pressure or flow reductions (if necessary) to prevent further damage or leakage. For part wall, non-leaking pipeline defects, it is usual practice to lower the operating pressure to 80% of that at which the defect was discovered/inflicted, until the defect has been assessed.

Further pressure reductions may be required, depending on the type of repair/rehabilitation that may have to be planned. For example, some composite repairs require pressure reductions of 50%. Similarly, recoating may require removal of the pipe from the trench, and the pressure may have to be reduced to zero.

3.2 Establish critical information for assessment

The fitness-for-purpose assessment, and the specifications of possible repair/rehabilitation methods, will need essential information. Some of the data needed are:

i. defect information: based on visual assessment, measurements and NDT,
ii. defect/failure history of line,
iii. pipe & weld material: pipe material grade, diameter, wall thickness, type of weld, coating, and pipe specification properties,
iv. construction, design & operating characteristics: design specification, statutory duties, design life, current operation, cyclic & hydrotest pressures (pre-service & operational), operating temperature, fluids, leakage rate, etc.,
v. pipeline configuration: defect locations, location of adjacent welds, bends and fittings, curvature in pipe, ovality of pipe, and
vi. consequences of failure (either reported or - if no failure yet - possible).

3.3 Location assessment

A location assessment will be needed to determine the requirements for, and availability of, repair/rehabilitation methods. The location of the damage or rehabilitation need, and access, will dictate the repair/rehabilitation methods. Damage can be located at various locations, each with differing access limitations: platform topsides, offshore riser, deeper water depth zone (>50m), intermediate water depth zone (20-50m), shallow water depth zone (<20m), inter-tidal zone, inshore zone, onshore, environmentally sensitive area, valve stations/pigging facilities/terminals, and road/river crossings.

4. STRATEGY STAGE 3 - ASSESSING DEFECTS AND DAMAGE IN A TRANSMISSION PIPELINE SYSTEM

4.1 Introduction

The reported defects can be assessed using fitness-for-purpose methods. However, the urgency of the analysis and any resulting repair depends upon:

i. defect severity: location, depth, length, orientation,
ii. financial/strategic value of pipeline,
iii. threat to environment & public relations,
iv. regulatory/legal/insurance considerations,
v. failure/further failures consequences.

4.2 Assessment methods

There are many documents and publications that assist pipeline operators to assess the significance of defects in pipelines. Most have their basis in Reference 12, and they are summarised in References 14 and 15.

Several papers at the present conference will cover these assessment methods, and therefore they are not covered in this paper. However, it should be noted that two new documents will soon be available to assess the significance of defects in pipelines. BSI PD 6493 (16) is being replaced and its update (BSI 7910) will contain guidelines on assessing defects in pipelines. API 579 - 'Recommended Practice for Fitness for Service', is a draft standard, expected to be published in 1999, and covers the assessment of defects in tanks, process plant and piping (17).

It should be noted that not all 'defects' are pipewall defects. Some will be structural anomalies (e.g. buckles or unsupported spans) that will require design/structural analysis.

4.3 Constructing defect assessment plots

Once you have decided on your assessment method (which will be governed by the defects in question), the basic equations are used to construct acceptance curves (18). These allow the operator to decide on which defects to accept without intervention, or which to repair or rehabilitate. These decisions will depend of the significance of the defects, and the cost model (e.g. Figure 4) being used to determine cost effectiveness.

5. STRATEGY STAGE 4 - REPAIR AND REHABILITATION

The last stage of a complete repair and rehabilitation programme is the actual repair and rehabilitation. The fitness for purpose assessments will have determined which defects to repair, and - if the defects/damage are extensive - a rehabilitation programme may have to be implemented.

5.1 Repair/Rehabilitation methods

There are many types of repair and rehabilitation methods, including: (i) Grinding, (ii) Weld Metal Deposition, (iii) Full Circumferential (Welded) Sleeves, (iv) Composite Reinforcement Sleeves, (v) Mechanical Clamps, (vi) Pipe Section or Pipeline Replacement, (vii) Upgrading Cathodic Protection, (viii) External (re)Coatings/Internal Liners, and (ix) Combined inspections/repairs/rehabilitation.

Most of these methods, particularly items i. to vii., are extensively covered in the literature, e.g. References 5, 10, and 19. Item (ix) is covered in the literature (e.g. (8)), therefore, only the internal linings and external recoating are summarised below.

5.2 Internal liners

Pipelines can be fitted with an internal liner to prevent further internal corrosion; however, these systems have limitations:

- pipeline has to be taken out of service and cleaned,
- liner can only be installed in short lengths (typically 800m),
- ends of the liner have to be sealed/joined to next section,
- there are concerns over use on sour service pipelines.

There are many possible systems at present, e.g.:

- Cured in-place - a thin reinforced textile liner, fixed by adhesive,
- Modified slip lining - polyethylene liner compressed & expanded,
- U-Process - polyethylene deformed to U shape, rolled in & expanded,
- High Density Polyethylene – but the installation method is not established.

5.3 External coatings

The following types of coating are commonly used onsite for the rehabilitation of onshore pipelines (20): coal tar enamel, cold applied laminate tapes, heat shrink materials, brush and spray applied two pack epoxy systems (both pure and tar modified), and brush and spray applied polyurethane systems (both pure and tar modified).

These new coatings are applied as follows:

COATING REMOVAL - mechanical beating, mechanised scraping, high pressure water jets. Water jetting is extensively used, and can remove coating (out of ditch) on a 24" pipeline at a rate of ~3m/min. In situ rates are lower because of extra care with handling the pipeline. Water jetting is better than sand jetting as it eliminates soluble salts in the corroded areas, which, if not removed, could initiate corrosion.

SURFACE PREPARATION - blast cleaning (manual/line travel) by steel shot or sand.

RECOATING - An American Gas Association study gave the following rehabilitation preferences (compared to new constructions):

Table 4. Coatings used in rehabilitation of landpipelines (21, 22)

	SMALL DIAMETER, % REHABILITATION (NEW)	**LARGE DIAMETER, % REHABILITATION** (NEW)
FBE	**0** (46)	**0** (63)
COAL TAR	**18** (13)	**17** (19)
TAPE	**46** (18)	**33** (7)
PRECOATED	**18** (0)	**25** (0)
ASPHALT ENAMEL	**0** (4)	**0** (7)
EXTRUDED PE	**0** (29)	**0** (4)
COAL TAR EPOXY	**18** (0)	**17** (0)

A report on a Canadian project (23) gives a production rate of between 1 and 2 km/day for recoating (epoxy urethane) a 36" pipeline, and a saving of 40% compared to replacing the pipeline. Mechanical wheel blasting and urethane coating will give you a rate of ~10m/min on a 30" line. However, tape is the most popular, if not favoured, rehabilitation. coating.

5.4 Lessons Learnt

The lessons learnt from one North American company during an onshore oil line rehabilitation (using urethane coatings) were (24):

1. detailed inspection is needed prior to rehabilitation,
2. detailed stress and fracture analysis is required to show that the pipeline could be recoated while 'loaded' (could be done in trench, but not out-of-ditch),
3. it was difficult to compensate for thermal expansion and locked in stresses when the pipe is exposed,
4. a safety programme essential,
5. the project was costly, but it had a slight advantage over replacement,
6. many improvements are needed to coating process,
7. site access is important, and...
8. expect the unexpected!

6. CONCLUSIONS

1. The repair and rehabilitation of a pipeline consists of four stages: inspection, response, assessment, and repair & rehabilitation.
2. Rehabilitation and repair is an engineering exercise, similar to a new construction, and should never be treated in isolation.
3. A repair and rehabilitation programme should be carefully costed, and the safest, most cost effective route selected.

REFERENCES

1. Anon., 'Interstate natural gas pipelines - Delivering energy safely', Interstate Natural Gas Association of America Report, USA, 1994.
2. Hopkins, P., 'Transmission pipelines: How to improve their integrity and prevent failures', 2nd Int. Conference. on Pipeline Technology, Ostende, Belgium, 1995.
3. Anon., '25 Years of CONCAWE Pipeline Incident Statistics', CONCAWE Vol. 6, No. 2, October 1997. (see http://www.concawe.be/html/vol62/statistics.htm).
4. Fadeeva, H. V., 'Environmental Aspects of Pipelining in Russia', 2nd International Onshore Pipelines Conference, Istanbul, 3-4 December 1998.
5. Curson, N., 'Options For Pipeline Replacement and Rehabilitation', Conference on 'Advances in Pipeline Technology '97', Dubai, IBC, September 1997.
6. Trefanenko, B. et al, 'Risk Assessment: an Integrity Management Tool', Pipeline Risk Assessment, Rehabilitation and Repair Conference, Houston, 1992.
7. Ketszeri, C., Storey, D. F., 'Optimising a Gas Pipeline Rehabilitation Program - a New Approach', 3rd International Conference on 'Pipeline Rehabilitation and Maintenance', Abu Dhabi, May 1999.

8. Taylor, S. A., 'Worldwide Rehabilitation Work is Undergoing Major Changes', Pipeline Industry, February 1994, p25.
9. DeHaven, T., 'Planning a Pipeline Rehabilitation Project', Pipeline and Utilities Construction, Vol 48, No 7, July 1993.
10. Hopkins, P., Bruton, D., 'Rehabilitation and Repair of Pipelines - Key Decisions', 'Rehabilitation: Piping and Infrastructure', University of Newcastle, March 1999.
11. Hopkins, P., Haswell, J., 'Practical Defect Assessment Methods for Application to UK Gas Transmission Pipelines', The Institute of Materials 2nd Griffith Conference, Sheffield, UK, September 1995.
12. Kiefner, J. F. et al., 'Failure Stress Levels of Flaws in Pressurised Cylinders', ASTM STP 536, pp 461-481, 1973.
13. Shannon, R. W. E., 1974, 'The Failure Behaviour of Linepipe Defects', Int. J Press Vessel & Piping, (2), pp 243-255.
14. Hopkins, P., Cosham, A., ''How to Assess Defects in Your Pipelines Using Fitness-for-Purpose Methods', 'Advances in Pipeline Technology '97' Conference, Dubai, IBC, Sept. 1997.
15. Hopkins, P., 'Ensuring the Safe Operation of Older Pipelines', Int. Pipeline and Offshore Contractors Association, 28th Convention, Acapulco, Mexico, Sept. 1994.
16. Anon., 'Guidance on Methods for the Derivation of Defect Acceptance Levels in Fusion Welds', BSI PD6493, British Standards Institution, London, 1991. (Update in 1999 to BSI 7910).
17. Anon., 'Recommended Practice for Fitness for Service', API 579 is a draft standard, expected to be published in 1999.
18. Hopkins, P., Lamb, M., 'Incorporating Intelligent Pigging Into Your Pipeline Integrity Management System', Onshore Pipelines Conference, Berlin, Germany, 8-9th December 1997.
19. Kiefner, J. F. et al 'Pipeline Repair Manual', American Gas Association, AGA Report PR-218-9307, 1993
20. Norman, D., Swinburne, R., 'Polyurethane and Epoxy coatings for the Rehabilitation and Repair of Pipelines', 3rd International Conference on 'Pipeline Rehabilitation and Maintenance', Abu Dhabi, May 1999.
21. Coates, A. C., Thomas E. B., 'Aging of Pipelines: Risk Assessment, Rehabilitation and Repair', Proc., 12th Offshore Mechanics & Arctic Engineering Conf., Glasgow, June 1993, Vol 3.
22. Taylor, S. A., 'Techniques for the Refurbishment of Oil and Gas Pipelines', Proc., UK Corrosion Conf., Manchester, Oct. 1991.
23. Alliston, B., Dzatko, J., 'Use of state of the Art SP-2888 RG Epoxy Urethane Coating for Pipeline Rehabilitation on TransCanada Pipelines and Great Lakes Transmission Pipeline System', 3rd International Conference on 'Pipeline Rehabilitation and Maintenance', Abu Dhabi, May 1999.
24. Pipeline and Gas Journal, October 1995.

C571/023/99

Recoating large diameter gas transmission pipelines in Western Canada

T D LEAHY
TransCanada PipeLines, Saskatchewan, Canada
M R PRIOR
Marine Pipeline Construction, Alberta, Canada
S A TAYLOR
CRC Evans Rehabilitation Systems, Houston, USA

ABSTRACT
Rehabilitation work on pipeline sections containing damaged or disbonded coating, corrosion and stress corrosion cracking has been–and continues to be–done manually. It is slow and labor intensive, usually with inconsistent results. Automated equipment for coating removal, surface preparation and coating application has been developed to provide superior performance and results while decreasing labor requirements. It is making rehabilitation of pipelines an increasingly attractive alternative to new construction.

1. INTRODUCTION

This paper examines an approach to maintaining the integrity of an aging pipeline system in Western Canada by recoating pipe sections to which the original coating has become damaged, disbonded or both. Experience has shown that this method of pipeline rehabilitation requires a close working relationship between the owner, prime contractor and the major rehabilitation equipment supplier. TransCanada PipeLines, Marine Pipeline Construction and CRC-Evans Pipeline International, the three primary partners in this ongoing project, present this paper.

2. SELECTING THE LINE TO BE REHABILITATED

2.1. The Goal: System Integrity

TransCanada PipeLines started to build its first Canadian pipeline in 1956. Since then, the Canadian mainline system has expanded to over 14,000 km of pipe in more than 3,000 km of right of way (ROW). This pipeline system can export 7.2 BSCD of natural gas from the fields in Alberta and Saskatchewan to Eastern Canada and the USA. In Western Canada, there are as many as seven lines running in parallel, in a single ROW (see Figure 1).

To ensure safe operations, TransCanada PipeLines developed a strategy for managing and maintaining the integrity of this system based on:

- Condition Monitoring. Acquiring information as to the condition of the pipeline.
- Risk Analysis. Knowing the likelihood and consequences of failure.
- Mitigation. Minimizing the risk by using a combination of approved pipeline maintenance techniques.
- Research. Refinements and improvements of the techniques for mitigation and risk analysis.

The strategy is applied to the subject line including its influence on all the other lines running parallel in the same ROW.

This paper concerns itself primarily with condition monitoring and mitigation. Condition monitoring is performed by running inline inspection tools, exposing and examining the pipe at discreet locations, analyzing cathodic protection (CP) data, and collecting data on the condition of the pipe and its coating system whenever the pipeline is exposed for any reason.

Mitigation techniques include hydrostatic testing, cathodic protection, routine maintenance, pipeline recoating and pipeline replacement.

Pipeline recoating is a long-term solution. Immediate or short-term threats to the integrity of the pipeline are usually addressed by the time recoating is considered.

2.2. The Problem: Corrosion

TransCanada PipeLines' current system of pipelines varies in age from one to forty-one years. The age factor, along with the different pipeline coating systems used in past times and the aggressive soil conditions present a challenge for operating this system safely while remaining competitive.

Soil conditions on the Canadian prairies are largely comprised of glacial till. This can take the form of unusually tenacious clay, which sticks tightly to the surface of the pipeline coating. The clay also expands and contracts, depending on its moisture content. Over time, the combined effect tends to tear the coating off the pipe. Additionally, the soils also vary in pH from 4 to 9 and are, therefore, very corrosive to exposed steel.

Pipeline construction over the last 40 years has seen the use of coating systems such as coal tar, asphalt, polyethylene tape and fusion-bond epoxy. The systems of particular concern are the asphalt and tape coated lines. Long sections of asphalt coated older lines no longer have coating on 60 – 70% of their surfaces. Middle aged lines, which are partly coated with polyethylene tape, have large sections of coating disbondment. Cathodic protection is the primary method of preventing corrosion on these sections. Tapecoat sections present even greater problems because the disbonded tape electrically screens the pipe from the cathodic protection system.

In the past, TransCanada PipeLines relied heavily on Cathodic Protection (CP) to prevent pipeline corrosion. For example, in one 23-km section of ROW there are 182 rectifiers putting out 2,500 amperes of current. In that same section, however, there are seven lines in parallel, all electrically common. This makes it very difficult to direct the CP only to where it is needed. In addition, high CP currents cause the fusion-bond epoxy coatings on the newer lines to disbond. To better control the CP system, TransCanada PipeLines is in the process of electrically isolating lines. CP is now considered as only one component in the overall strategy of mitigating the risk of corrosion causing a pipeline failure.

2.3. Selecting the Line and Section to be Recoated

TransCanada PipeLines' Pipeline Integrity Department collects and analyzes pipeline condition data on an ongoing basis. This data, taken from inline inspections and CP surveys, forms the basis for medium- and long-term mitigation strategies. From this analysis comes the decision as to which of the potential solutions best suits the circumstances based on integrity and economics. Table 1 demonstrates how this data is used to select the lines to be recoated and how the chosen lines are prioritized.

Table 1
Typical Evaluation Data Used for Recoating Decision

Location	Avg. No. defects per 500 m	Coating Type	CP Levels	Percent increase, 5 yr.	Ranking	Comments
A	117	Asphalt	-700mV in '90, -550mV in '95, -750mV in '98	266%	N/A	Add CP
B	442	Asphalt	-650mV in '90, -400mV in '95, -500mv in '98	266%	1	Recoat in 2000
C	883	Asphalt	-900mV in 89, -750mV in '98, P/S up and down	101%	2	Recoat in 2001
D	584	Asphalt	Very low potentials in '90, -1200mVs in 96, +200mV polarization, A few anomalies	196%	1	Recoat in 2000
E	579	Asphalt/ Tape	-850mV to –1500mV in '95, -500mV to –1000mVin '98, P/S up and down, A lot of anomalies	3%	2	Recoat in 2001

P/S = Pipe to Soil potential on/off readings

Notes:

1) Location "A" shows a CP level of –750mV and a low number of anomalies. It was decided that adding more CP facilities was the most cost-effective solution.

2) Locations "B" and "D" show a ranking of 1 because even after increasing the CP levels by 266% and 196%, the CP level cannot be raised to the prescribed levels. Also, the section is partly coated with polyethylene tape.

3) Locations "C" and "E" show a ranking of 2 because although the general CP levels are not so bad, the pipe-to-soil readings show localized anomalies. We also know that the electrical current is very high in these sections.

The process is one of looking for a combination of the following existing factors:

- soil type that has a propensity for corrosion,
- current CP levels, their effectiveness and their effect on the adjacent lines,
- coating type that is either polyethylene tape, asphalt or coal tar,
- coating condition, and
- corrosion defect indications (features) per unit of length.

An important external factor in the decision to recoat is the type of terrain in which the section of pipe is buried. If the pipe is laid in rocky, hilly, sandy or wet ground, the cost of exposing, recoating, and reburying the line could be prohibitive, as compared to the cost of laying new pipe. Table 2 shows recoating costs for TransCanada PipeLines over the past three years:

Table 2
Costs of Mainline Recoating Program

	Length (Km)	Size (mm)	Old Coating	Total Cost ($)	Cost /Km ($)
1996	1.6	864	Asphalt	$1.2 MM	$750,000
1997	14.5	864	Asphalt	$8.7 MM	$600,000
1998	26.2	864	Asphalt	$15.80 MM	$603,000

It is estimated that the cost of replacing NPS34 pipe in this part of Western Canada is approximately $977 thousand per km. Therefore, in 1998 the saving by rehabilitating the pipe by recoating is estimated at 38%. (The 1998 cost per km does not show a decrease over the 1997 cost because of that year's unusually wet summer.)

3. THE CONSTRUCTION PROCESS

Rehabilitation projects are large and complicated undertakings requiring major pipeline contractors for successful execution.[1, 2] Marine Pipeline Construction is the largest pipeline construction company in Canada. Engaged primarily in the construction of new pipelines, it is also actively involved in the maintenance and rehabilitation of existing pipelines, including investigative inspection, retesting, pipe removal, recoating and replacement.

Marine has played an integral role in TransCanada PipeLines' rehabilitation efforts in the past three years. As each project proved successful, subsequent ones have increased in scope and complexity.

3.1. Exposing the Pipeline for Reconditioning

Preparation of a ROW that is both accessible to heavy construction equipment and protective of adjacent land is critical. The area of work is mostly prime farmland, but there are sections of environmentally sensitive wetlands and other locations where rare flora and fauna unique to the Western Canada prairie can be found. The region is also a major flyway for migrating waterfowl. And, virtually all land is privately owned and must be treated with care and concern.

Working closely with environmentalists, the construction team develops a detailed grading plan as to how ROW preparation will be carried out. Removal and preservation of the active organic soil layer (topsoil) is necessary so that heavy construction may proceed without environmental damage. Following topsoil removal, the subsoil is graded where necessary to make the contour of the land suitable for the safe operation of heavy equipment.

Creeks and wetlands are the most sensitive areas with regard to ROW preparation. Due to the tremendous variety of plant and animal life in the vicinity of these areas, water crossings must be made with great care using environmentally protective techniques such as fluming, dam and pump techniques, and open cut methods. Extensive use is made of geo-textiles for filtration and ground-bearing improvement.

Following preparation of the ROW, the existing pipeline is excavated and prepared for inspection. The lines here are typically covered by 1.2 meters of subsoil. Removal is accomplished with the use of a wheel ditcher set to excavate 1.1 meters of subsoil. This equipment is designed to excavate much deeper. Consequently, extreme caution must be exercised in removing the maximum possible overburden without contacting the pipe surface as that usually results in an expensive pipe replacement.

The pipe is then lifted from the ditch and set on skids. This work is done by sidebooms using rolling cradles under the pipe to successively lift it from the ditch. The cradles are fitted with scraper devices to remove remaining dirt stuck to the pipe surface. Then the old ditch is backfilled and the pipe reset over it in order to create sufficient workroom on the ROW.

Labor crews remove remaining dirt and loose pipeline coating materials to locate the previously identified areas of significant defects. These pipe sections are blasted sufficient to confirm the initial defect assessment. Defective pipe sections are cut out and replaced.

3.2. Removal of Existing Coatings

The objectives of the coating removal operation are twofold: First, to remove the old coating material and corrosion deposits to allow for visual inspection of the underlying steel surface.[3, 4] And, second, to remove water-soluble contaminants that could create a site for corrosion under the new coating.[5]

The HydroKleaner™, shown in Photo 1, sits on the pipeline and houses the high-pressure swivels. At pressures to 1,400 bar (20,000 psi), water jets can remove existing coating, corrosion products and soluble salts, leaving only the primer stain. High-pressure water is supplied by the pump unit at a rate of 189 Lpm (50 gpm) and 1,400 bar (20,000 psi). A nurse truck refills the water tanker reservoir while a sideboom provides travel.

Water used for coating removal operations must be of potable grade and free of contaminants such as chlorides and anaerobic bacteria.[6] Transport units must also be contaminant free. The following water analysis is recommended:

Maximum Particle size:	<10 microns
Suspended Solids:	<50 parts per million
Dissolved solids:	<50 parts per million
Conductivity:	30 to 50 microhm/cm

Removed coating material and water are contained within the HydroKleaner shroudand directed to a shredder which breaks up large coating pieces. A vacuum line attached to the discharge spout on the shredder directs collected material and water to a *Super Sucker*™ truck. A minimum of two trucks is recommended for continuous operation.

The filled *Super Sucker* truck discharges its load into a lined pit. Water and fine particles of coating material are pumped from the pit to a filter trailer, where the fine particles are separated from the water by 10-micron filter cartridges. Clean water is pumped to a water storage tank and, after analysis, dumped to earth. The remaining coating material and filter bags can be taken to an acceptable dumpsite for disposal.

Average linear production rates of 3.0 meters per minute (9.8 ft/min) were achieved on 34-inch pipe with well-adhered coal tar enamel (CTE) coatings up to 6 mm (0.25 inch) thick. Line travel rates to 7.0 meters per minute (23 ft/min) were obtained on poorly bonded coating.

Cold applied tape is very difficult to remove—in particular the butyl, the adhesive left after tape removal. At pressures of 1,400 bar (20,000 psi), productions rates of only 1.04 meters per minute (3.4 ft/min) were obtained on multi-layer tape systems. This is one third the minimum cleaning rate expected for CTE.

At pressures of 2,400 bar (32,000 psi), the water jets have improved cleaning capabilities. They are able to remove primer stain and flash rust and, more importantly, they can remove cold applied tapes at significantly faster rates. Initial tests show that four-layer tape systems can be removed at production rates of 2.4 linear meters per minute (7.9 ft/min) on 34-inch pipe. This is 2.3 times faster than water jets at 1,400 bar.

3.3. Pipeline Inspection and Repair

The goal is to ensure that before the pipe is recoated and buried, all defects are discovered and repaired or removed. The pipeline section is repaired in accordance with CSA Z662-96.[7]

Based on the results of previous inspection using a Magnetic Flux Leakage (MFL) tool, short sections of the pipe with severe or a high density of corrosion features will be cut out and replaced with pre-tested pipe. Predetermining pipe cutouts allows the prime contractor to better plan the work; it also reduces out-of -service time.

Corrosion pits are evaluated using the RSTRENG method, and the burst pressure is calculated.[8] If the result is less than 100% SMYS pressure, the defect is cut out. Where a residence is located within 200 meters, 105% SMYS pressure is used as the criteria.

At discreet locations along the exposed section, the pipe is blast-cleaned by hand and inspected for SCC using Magnetic Particle Inspection (MPI) techniques. Appropriate repairs are conducted. The locations chosen are based on soil model analysis and the coating system. Inspection and repairs are also made for all other types of defects normally found in pipeline of this vintage, including gouges, arc burns, dents, weld defects, etc.

3.4. Surface Preparation and Recoating

Surface preparation is the single most important factor in the success or failure of a protective coating.[9, 10] Surfaces, including the anchor pattern, are prepared as specified by the coating manufacturer or pipeline owner. Quality assurance is achieved through routine measurements of anchor pattern depths using Replica Tape and a Spring Micrometer.

A mechanical wheel blast system is used to prepare the pipe surface.[11] The Line Travel Mechanical Blast (LTMB), shown in Photo 2, consists of eight Rotoblast wheels that hurl abrasive media at the pipe surface from within the enclosed system. The self-cradling LTMB system can prepare the pipe surface to any commonly accepted standard in one pass.

Its equipment sled includes a dust collector to remove dust, rust and foreign materials from the line unit. A diesel-driven generator provides electrical power. An air compressor provides air to clean the shot and grit from the pipe before it exits the LTMB. A central control panel allows the operator complete operational control.

The size and ratio of the shot and grit materials determine the surface preparation and profile obtained.[12] Initially, a mixture of shot and grit is required to establish proper abrasive flow through the system. During normal operation, replenishment abrasive is usually 100 % grit.

Rigorous attention is given to ensure uniform application of coating to pipeline steels. The line travel coating system (CPCO), shown in Photo 3, uses three spray guns to provide uniform 360-degree coverage. Coating materials are metered, pumped through static mixing tubes, and then sprayed on to the pipe at pressures up to 263 bar (3,500 psi). A programmable logic controller (PLC) controls all functions of the coating system.

The CPCO equipment skid contains a control panel and MIS-metering system. It also includes heated material component tanks, transfer pumps and breathing air. This unit permits the contractor to load the coating material at a central depot for transport to the coating unit as required throughout the day. Applying liquid coating materials with airless spray technology results in some over-spray. Therefore, equipment in the vicinity should be coated with a "release" material to permit cleanup at the end of the project.

The liquid epoxy coating applied requires time to cure. The time can vary from 20 minutes on a hot day to several hours on a cold day. During the curing time, the coating is very susceptible to insect damage. Normal repair procedure is to allow the coating to cure to a tack-free condition. A crew then manually removes the insects from the coating. A hand spray unit makes coating repairs at the defect site. Holidays and other coating defects are repaired to the same exacting standards used to apply the initial application.

When the new coating is at least 80% cured, Dry Film Thickness (DFT) measurements are taken in accordance with SSPC Recommended Practice PA-2.[13] Then the coated line is inspected using electrostatic holiday test equipment according to the coating manufacturer's recommendations. All defects are marked and repaired using "brush-grade" material.

3.5. Pipeline Reinstallation

After the pipeline has been cleaned, inspected, repaired and recoated, it is ready for reinstallation. During the recoating process, the pipe was moved away from the ditchline and reset on skids. The original ditch is now re-excavated and prepared for installation of the recoated pipeline.

The pipeline is cradled by sidebooms over the ditchline, and a final check for holidays is made before lowering the pipe into the ditch. The ditch is then backfilled, and the first stage of the ROW cleanup is made. Road crossings, rail crossings and creek and river crossings are installed separately using environmentally sound methods.

Hydrostatic testing is performed to confirm the integrity of the line and to conform to CSA Z662-96 testing requirements.[7] Cleaning pigs first remove traces of oils and other environmentally harmful materials, which are collected and disposed of. Then the pipeline is filled with water and subjected to a hydrostatic test at a minimum of 1.25 times its maximum allowable operating pressure. The recoated pipeline section is tied back into the adjacent sections. After the final tie-in welds are X-rayed, the section is filled with natural gas. The pipeline is now placed in service, its integrity comparable to a newly constructed pipeline. From start to finish, the pipeline section is normally out of service approximately 30 days, plus 1.2 days per kilometer of pipe recoated.

3.6. Scheduling and Cost Control

Key to the operation for all parties is the ability to maintain a set schedule that minimizes out-of-service time. The objective, after all, is to rehabilitate pipeline sections at a measurably lower cost than that of new construction.

Before starting the work, the owner and prime contractor review all elements of the project to ensure all possible contingencies and potential construction efficiencies are addressed.

Extensive planning sessions are held with all involved parties. A project schedule, called the Baseline Schedule, is developed and includes each material supply and construction operation. It is the standard against which progress will be measured. A successful project requires schedule input from all participants. Likewise, all participants must show ownership of the Baseline Schedule.

When necessary, additional equipment and manpower are moved to critical parts of the work in order to maintain planned production rates. Float and time allowances for contingencies are not given up without a thorough analysis as to the need.

Pipeline construction is very weather dependent, rehabilitation programs even more so. Pipe temperature, dew point and relative humidity are critical factors in addition to the obvious impact of wind, rain and insects.[14] Excessive humidity can cause oxidation of the steel surface immediately after surface preparation. Ambient conditions must be correct for the coating material to properly cure. Waiting for acceptable conditions can significantly delay daily starting times. Fortunately, the long summer days and predominantly sunny weather of the Prairie Provinces can be used to recover some lost time.

Cost control is vital to project success and starts with an approved and accepted detailed cost estimate and schedule. As it occurs, each cost element is compared to the original estimate; major cost reports are prepared weekly and used to project completion costs against original budget estimates. These discussed methods have proved successful, as recoating projects undertaken to date have been completed within their original budgets.

SUMMARY

After three years of joint efforts, TransCanada PipeLines, Marine Pipeline and CRC-Evans Pipeline International have indeed demonstrated that large diameter pipelines can be recoated and replaced in service at a cost far less than that of new construction. A successful recoating project is dependent on a close working relationship among the primary partners and their ability to plan the work and solve major problems in advance. As experience is gained, equipment and work methods will be refined to reduce recoating costs even further.

REFERENCES

1. W. Kresic, "Urethane Coatings Rehabilitate Large Crude Oil Pipeline," Pipeline & Gas Journal, October 1995, pp.41-48.
2. Y.M. Ireland, A. Petrusev and A. Lopez, "Field Application of Rehabilitation Coatings for Pipelines," presented to 1999 Western Conference (Calgary, Alberta: NACE 1999).
3. L.M. Frenzel, "Application of High Pressure Water Jetting in Corrosion Control," SSPC Annual Symposium (Pittsburgh, PA: SSPC, 1985) pp. 164-184.
4. L.M. Frenzel, "Evaluation of 20,000 psi Water Jetting for Surface Preparation of Steel Prior to Coating," SSPC Annual Symposium (Pittsburgh, PA: SSPC, 1983).
5. E.D. McCrory, MP 26, 3(1987): pp. 49-54
6. D.G. Weldon, "Salts: Their Detection and Influence on Coating Performance," in Proceedings of the SSPC Annual Symposium (Pittsburgh, PA: SSPC, 1985).
7. "Oil and Gas Pipeline Systems," Section 10.8, "Evaluation of Imperfections and Repair of Piping Containing Defects."
8. *RESTRENG for Windows,* Technical Toolboxes, Inc. 3801 Kirby Drive, Suite 340, Houston, TX 77098
9. D.A. Bayliss, "Surface Preparation–The State of the Art," in Proceedings of the SSPC Annual Symposium (Pittsburgh, PA, 1985), pp. 220-221.
10. T.J. Pfaff, "Requirements for External Pipeline Refurbishing Coatings, presented to the 1989 Pipeline Rehabilitation Seminar, Houston, TX.
11. A.W. Mallory, "Mechanical Surface Preparation: Centrifugal Blast Cleaning," Good Painting Practice, Steel Structures Painting Manual, Vol. I, 2nd ed. (Pittsburgh, PA SSPC, 1982) pp. 22-35.
12. E.A. Borch, "Metallic Abrasives," Good Painting Practice, Steel Structures Painting Manual, Vol. I, 2nd ed. (Pittsburgh, PA SSPC, 1982) p.35
13. "Paint Application Specification No. 2," Systems and Specifications, Steel Structures Painting Manual, Vol. II, 7th ed. (Pittsburgh, PA SSPC, 1982) pp. 407-409.
14. A. Lopez, Y. Ireland and P. Powers, "Coating Trial and Selection of Rehabilitation coatings for Large Diameter Pipe," presented to 1999 Western Conference (Calgary, Alberta: NACE 1999).

Figure 1
TransCanada PipeLines System Map

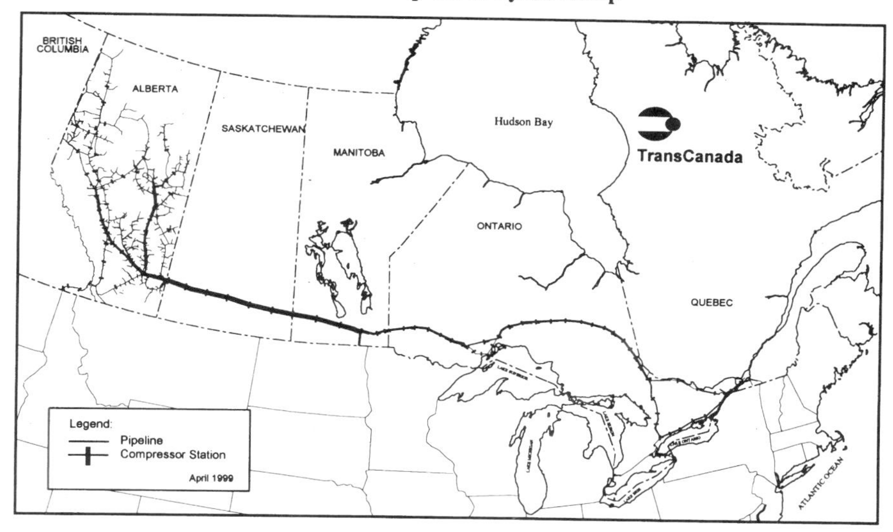

Picture 1
The HydroKleaner™

Picture 2
The Line Travel Mechanical Blast Machine (LTMB™)

Picture 3
The Line Travel Plural Component Coating Machine (CPCO™)

C571/042/99

Decommissioning and abandonment of onshore pipelines

M BROWN
Transco Engineering Services, Newcastle-upon-Tyne, UK

SYNOPSIS

The UK Pipelines Safety Regulations require the pipeline operator to ensure that a pipeline, once decommissioned, does not become a source of danger. Elsewhere, Regulatory Authorities require consideration of the possible issues which pipeline decommissioning may raise. This paper considers current practice with respect to pipeline decommissioning and abandonment, and sets out the range of factors which the operator must take into account.

1.0 INTRODUCTION

For the purposes of this paper: “decommissioning” is the removal of a pipeline from operational service, retaining the pipeline in a condition such that recommissioning would be possible; “abandonment” is a permanent removal from service, recommissioning is not intended and in most cases would not be possible.

Due to pipeline population ageing, gas and oil field depletion and a range of possible other changes in infrastructure, the issue of pipelines which no longer have useful economic lives will inevitably rise. At present there is no uniform opinion as to how to deal with this rising population; some pipeline codes do mention decommissioning, but this is usually in very general terms. In recognition of the low level of research and review in this field, a European Gas Research Group working committee, GERG PC 2.32, has reviewed European practice when abandoning pipelines, and is in process of commissioning research to develop a consensus methodology within Europe. This work will be published at a later date.

This paper considers the relevant issues and options and is intended to facilitate discussion on appropriate methodologies.

2.0 DECOMMISSIONING AND ABANDONMENT

2.1 General

Where the pipeline is to be taken out of use on a temporary basis, it is common practice to purge the pipeline and fill it with nitrogen, all normal maintenance activities (checking of cathodic protection etc) being maintained. In these circumstances, since the pipeline integrity is to be maintained at the in-service level, no special cleaning of the pipeline is required
If the decommissioning is to be permanent, and the pipeline is to be abandoned, other issues need to be considered.

Regulation 14 Decommissioning, of the UK Pipelines Safety Regulations (1), states "*The operator shall ensure that a pipeline which has ceased to be used for the conveyance of any fluid is left in a safe condition*".

The Guidance on Regulations (2), expands on this and requires that the pipeline be either removed or else left in a safe condition.

IGE / TD / 1 (3) (TD/1) which is the set of recommendations on design, construction operation etc of gas transmission pipelines (in the UK this is above 7 bar pipelines), most widely used in the UK, differentiates between temporary decommissioning and permanent decommissioning (abandonment) of a pipeline. Abandonment, which is a permanent removal of the pipeline from service, requires additional consideration.

2.2 Cleaning

The cleanliness of the pipeline is a key issue, whether the pipeline is to be abandoned and left in place, or is to be removed. For the pipeline that is to be removed, the cleaning must be such as to ensure that no harm to persons or to the environment will occur during the cutting up and removal process. For a pipeline left in-situ, water ingress following eventual loss of integrity due to the corrosion process and subsequent transport of any contaminants by water flow must be taken into account.

Cleaning of pipelines is a specialised area and the pipeline operator should hold detailed discussions with the contractor selected to carry out the cleaning process, to ensure that the high level of cleanliness required for an abandoned pipeline, can and will, be met. Some inspections of the internal condition of the pipeline following the cleaning process should be planned to confirm that the required standards have been achieved.

When planning the cleaning, the operating history records should be reviewed to establish the type of contamination which would be expected. Multiple runs with trains of cleaning pigs, perhaps pushing slugs of liquid solvent through the pipeline, may be required to meet abandonment cleanliness standards.

It should be mentioned that there is some suggestion that in non-sensitive areas a very high level of cleanliness is not essential since the surrounding soil will itself quickly filter out contaminants. With the current emphasis in the UK on environmental issues, it is unlikely that this approach would be acceptable in the UK legislative framework.

The major abandonment choices (Figure 1) are now summarised.

2.3 Removal

Removal is recommended for above ground sections of pipeline by TD/1; however it is recognised that economic considerations mean that removal will be limited to short underground sections, probably in areas of particular sensitivity / difficulty. Typical areas where removal may be given particular consideration would be Sites of Special Scientific Interest, water collecting areas, and major crossings of rivers, where there is a risk of the pipeline becoming uncovered. The level of disruption which any removal project would cause, and the benefit over other methods of abandonment, would also require to be taken into account. Railways, for example, require that the track remains very level to allow the train to operate safely. There is then a risk to the safety of the railway should the ground subside following removal of the pipeline. One of the other abandonment options, see below, may be preferred in this case.

Smaller diameter pipelines can be pulled out of the ground by large excavating machines, with only limited amounts of digging being required. Larger diameter pipelines would require a trench to be opened above the pipeline and the pipeline to be cut into lengths suitable for transport away from the site for final disposal. In many cases, particularly for older and lower pressure pipelines, the pipeline may be in a location where it is very difficult to remove safely, e.g. in a suburban area.

Figure 2 gives a summary of pipeline abandonment by removal.

2.4 Abandonment in-situ

2.4.1 Problems

There are a range of problems which may occur with in situ abandonment. A range of the problems which a failing, abandoned pipeline could cause is listed below.

i) Environmental impact.
It may not be possible to completely clean the pipeline. Assuming corrosion eventually perforates the pipeline, water will then be able to enter the pipeline and may wash residual pollutants out of the pipeline and into the surrounding area. This may be of particular concern near watercourses, where fish stocks etc. could be affected, or where groundwater is used to supply reservoirs. Current legislation in the UK would designate the abandoned pipeline as an environmental risk under the Environmental Protection Act, and the land would be designated as contaminated ground (5, 6).

ii) Drainage issues.
A pipeline perforated by corrosion could act as an unwelcome new drain or watercourse causing flooding or loss of water which could affect the agriculture use of the land. In extreme cases, ground stability could be affected by severe flooding with consequent problems for roads and buildings.

iii) Ground settlement.
If structural integrity is lost, the pipeline could collapse under the weight of ground loading, with subsequent ground settlement above the pipeline. This is likely to be many years following abandonment and would occur much later than the initial perforation of the pipeline due to corrosion. Collapse could cause serious problems if the pipeline runs below a railway line or a major road (depending on the method of construction) and also has an impact with respect to land use above the pipeline e.g. building subsidence. Ground settlement can also be an issue when a larger diameter pipeline is removed.

An estimate of the time taken to perforate a pipeline due to corrosion, following abandonment was determined in the Pipeline Abandonment Discussion Paper (4) commissioned by a Committee of Canadian interest groups. Using the example of a 323.9mm diameter pipeline, in soils typical of those found in Alberta Canada, it was estimated that perforation would occur between 13 and 123 years following the turning off of the cathodic protection system. Structural collapse would take significantly longer than this, since much more extensive corrosion would be required before collapse occured.

2.4.2 Mitigation of Abandonment Problems

1) Maintain corrosion protection
The simplest option is to maintain, so far as is possible, all of the activities directed to maintaining the pipeline integrity. The pipeline could be purged to inert gas, usually nitrogen, but with the cathodic protection (CP) system left operational, with periodic checking / auditing of the CP system being carried out as would be the case for a live pipeline. It could be argued that the purge operation is not required if the CP system is to be maintained since integrity will also be maintained. The cost and long term commitment this option would require should not be underestimated. The abandoned pipeline section should be isolated from the remainder of the pipeline, and an independent CP system installed. For short lengths, a sacrificial anode system is likely to be most cost effective.

2) Grout the pipeline
Completely filling the pipeline with grout / concrete will prevent the pipeline being used as a water course and also remove any possibility of structural collapse. However, for long lengths of pipeline this is not a practical option. This measure would normally be only considered in specific areas (for example a railway crossing) where any ground settlement could have serious safety implications. To ensure that the grout does fill the pipeline up to the top of the diameter, an appropriate arrangement of plugs and vents will be required. This approach leaves a permanent concrete structure in place.

3) Sectioning
More practically, the pipeline can be sectioned. Each section may be capped by a steel plate or else plugged by concrete or other impermeable material e.g. polyurethane foam (4), which will also adhere well to the pipe wall, to ensure a gas tight seal. If an inert gas is to be used to pressurise the section, consideration should be given to periodic leak testing, or to maintaining a facility to check the pressure in the section, to confirm continued integrity .

Figure 3 gives a summary of abandonment in situ, with mitigation measures.

2.5 Re-use

Before abandonment, care should be taken by the operator to ensure that neither they, nor any other pipeline operator has an alternative use for the pipeline. For example, in 1992 the then British Gas - Wales Region purchased a redundant Shell "oil" pipeline to provide additional gas storage capacity to North Wales. The purchase of this pipeline also allowed the replacement of much original North Wales pipelines, originally constructed in the 1950's. This 36" (900mm) diameter pipeline, which is approximately 122 kilometres long, is laid predominately across country from Stanlow in Cheshire to Rhosgoch on the island of Anglesey, just off the North West coast of North Wales.

Significant work had to be undertaken to confirm the suitability of the pipeline for gas duties and some additional works were required; however, the final total costs were far less than those which would have been incurred should a new pipeline have been laid.

More innovative uses suggested for abandoned pipeline sections have included conduits for electrical or optic fibre cables, particularly in difficult construction areas such as river crossings.

3.0 RECORDS

As a minimum, the pipeline operator should keep a record of: the precautions taken to ensure that the pipeline abandonment has been carried out safely; the abandonment methods and procedures; the location and other pipeline details (diameter, wall thickness); and any monitoring details. The relevant land owner should be made aware of the abandonment of the pipeline and fully informed of any mitigation measures taken.

4.0 THE ABANDONMENT PLAN

From the above, it is clear that abandoning a pipeline is an engineering problem which may be deceptively complex. Errors made during the abandonment process may have serious consequences, and could either be immediately apparent or else take several decades to appear.

For these reasons it is very important to ensure that all of the possible problem areas have been considered and dealt with.

Figures 2 to 4 give simple abandonment plans, depending on which route has been selected.

5.0 CONCLUSIONS

There are two major methods of pipeline abandonment: removal and in-situ abandonment.

Pipeline removal is likely to be the least favoured abandonment option in most circumstances, due to economic, environmental and technical considerations.

It is likely that when abandoning a pipeline, several of the different options raised above (removal, plugging, fully filling with grout etc.) may be required, depending on local conditions.

Re use is an alternative option, however the circumstances when this is feasible are likely to be rare.

Onshore pipeline abandonment is not well reported and researched. In recognition of this, work is now ongoing to establish "Guidelines for Best Practice" within Europe. Although this work is specifically directed towards natural gas pipelines it is likely that there will be much in common with all onshore pipeline abandonment.

REFERENCES

1. "The Pipelines Safety Regulations 1996", SI 825, HMSO Books.
2. "A Guide to the Pipelines Safety Regulations 1996", Guidance on Regulations, Health & Safety Executive, HMSO Books.
3. "Steel pipelines for high pressure gas transmission", IGE/TD/1 Edition 3, 1993, The Institution of Gas Engineers.
4. "Pipeline Abandonment, A Discussion Paper on Technical and Environmental Issues", Prepared for the Pipeline Abandonment Steering Committee" (comprised of representatives from the Canadian Association of Petroleum Producers, the Canadian Energy Pipeline Association, the Alberta Energy and Utilities Board, and the National Energy Board) November 1996.
5. "Pipelines, a worm's eye view", Julian Barnett and Michael Jordin, Transco Pipeline System Development Team, Transco in-house publication
6. "The Environmental Protection Act 1990", HMSO.

ACKNOWLEDGEMENTS

The Author would like to thank Transco for permission to publish this paper and to thank colleagues within the pipeline industry for their useful suggestions and comments. The views and opinions given within this paper are those of the Author, based on a general review of abandonment practice worldwide, and do not necessarily represent policies within Transco.

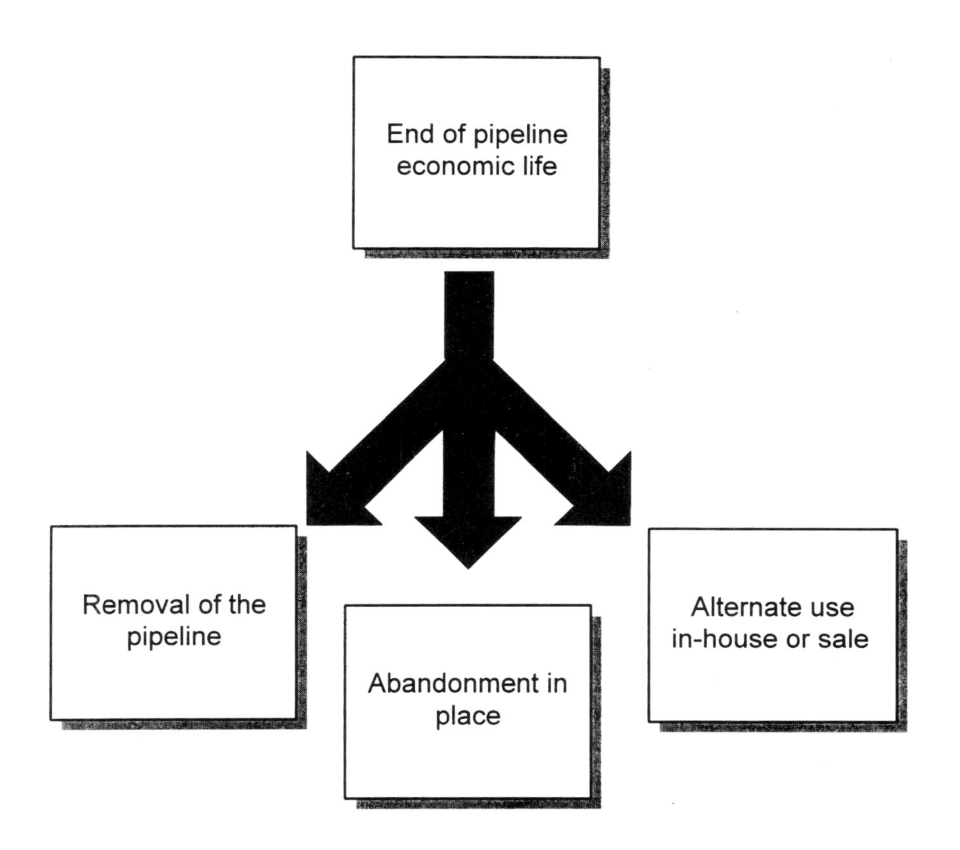

Figure 1 End of life options

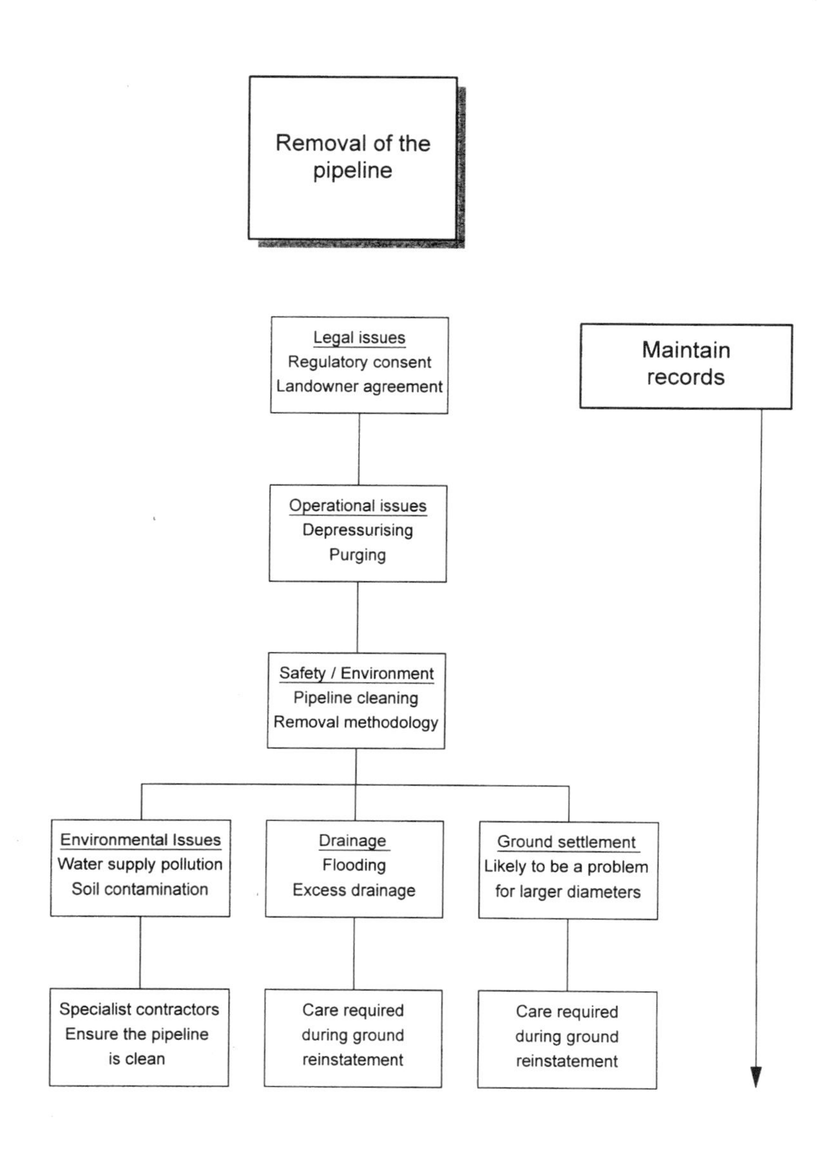

Figure 2 Removal of the pipeline

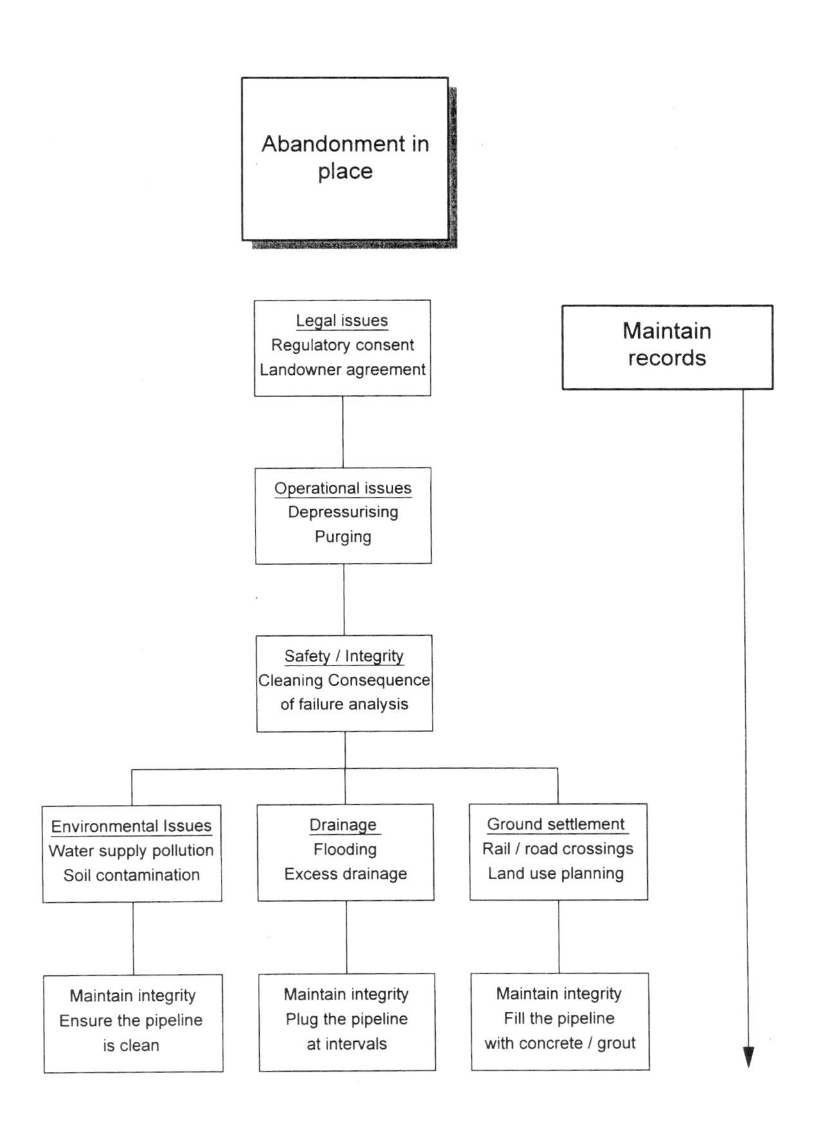

Figure 3 Abandonment in place

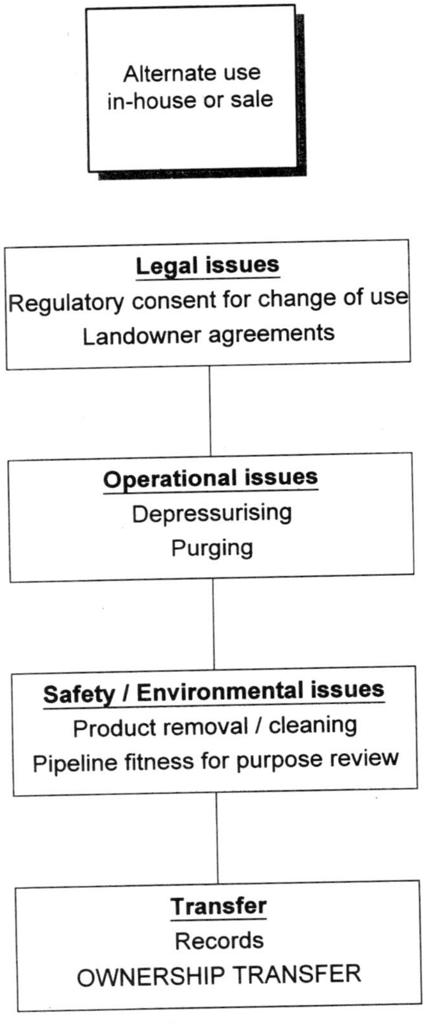

Alternate use
in-house or sale

Legal issues
Regulatory consent for change of use
Landowner agreements

Operational issues
Depressurising
Purging

Safety / Environmental issues
Product removal / cleaning
Pipeline fitness for purpose review

Transfer
Records
OWNERSHIP TRANSFER

Figure 4 Change of use or sale

Low Pressure:
Asset Management

C571/004/99

Condition assessment and its role in water pipeline asset management

S J DANIELS
Northumbrian Lyonnaise Technology and Research Centre, Newcastle-upon-Tyne, UK

1 SYNOPSIS

It is becoming increasingly important that water distribution companies around the world gain a better understanding of the current condition and performance of their underground assets. This increased knowledge will enable companies to better identify the risks they face, prioritise the needs for investment and to determine the most appropriate remedial action.

Buried pipes, valves and fittings can account for up to 70% of the total value of Water Company assets. Although catastrophic pipeline failure in a water supply network may not have the same safety implications for the operator as it might for a gas supply network; it can still result in loss of supply, major damage to public property and pose a severe local flooding risk.

To fully assess the condition of underground assets it will be necessary for water companies and their operators to undertake a detailed programme of inspection and diagnosis. This should build on the current practice of using indicators such as; failure frequencies, leakage history, ground conditions and pipe age to determine network condition.

2 INTRODUCTION

There are 10 Water & Sewage companies and 16 Water only companies in England & Wales who operate water supply networks totalling 325,000 km in length[1]. Between them the 26 individual water supply networks deliver into supply an average 15,683 million litres of water per day[2]. In the period 1991-98 more than 10% or, 33,000 km of water mains in the United Kingdom were relined or replaced. During 1997-98 alone water companies replaced or relined 5,870 km of water main.

There are a number of existing and developing condition assessment tools available to the operators of water and low-pressure gas networks. Using these techniques, it is possible not only to assess cracks, wall thickness, internal and external metal loss associated with corrosion, but also mechanical damage and manufacturing faults. This should allow the identification of locations requiring local repair and also the prioritisation of pipes for replacement or rehabilitation. These tools can be classified and will be discussed further under the following titles:

- Sample assessment
- CCTV inspection
- Metal loss monitoring

3 REGULATION & LEGISLATION

The United Kingdom water companies operate within a legislative framework set out in the Water Industry Act 1991[3]. The licence under which water companies operate requires them to produce and maintain an underground asset management plan (AMP). Through the legislation contained in the Act[3], the government has a number of measures designed to protect customers and the environment, these are overseen by three appointed regulators:

Ofwat (The Office of Water Services) – Supports the Director General of Water Services, whose role it is to monitor and audit the performance of water companies in relation to the licence, in terms of levels of service, charges and capital investment.

DWI (Drinking Water Inspectorate) – Ensures that water companies comply with their statutory duty to supply wholesome water.

EA (Environment Agency) – Ensures that the abstraction of water and its subsequent discharge is adequately managed to protect the environment.

Both Ofwat and the DWI have measures for monitoring Levels of Service to customers. In the supply of water these include:

- DG2 – Inadequate pressure (Ofwat)
- DG3 – Supply interruptions (Ofwat)
- DG4 – Restrictions on use of water (Ofwat)
- Mains bursts disrupting service (Ofwat)
- Water quality compliance failures (DWI)

Measures monitored by the respective regulators can have an impact on management of water company buried assets. For example, the value placed by customers on aesthetic water quality would justify the cost of mains relining operations to reduce incidence of discoloured water. Whereas, expenditure on issues to reduce leakage levels may have less value to customers, since there is no visible benefit.

4 INFRASTRUCTURE REHABILITATION

The overall objective of a water company's rehabilitation programme is to increase network reliability and efficiency thus allowing it to maintain the required service levels to customers. In doing this a company will aim to:

- Minimise the number of bursts disrupting service to customers
- Ensure that unplanned interruptions to supplies are reduced

- Reduce the physical losses
- Improve customer standards of service

4.1 Rehabilitation techniques

Techniques for the rehabilitation of water mains fall into two categories: renewal and relining. Renewal refers to those used as structural refurbishment techniques; these do not rely in any way on the structural integrity of the old main that they replace. Relining techniques are primarily designed to improve poor water quality associated with the internal corrosion of cast iron and steel pipelines and are generally not a structural solution, but rely on the host pipe to maintain structural integrity.

The decision whether to renew or reline must be based upon both technical and economic factors. Factors influencing the decision include:

- Structural condition of the existing main, Figure 1
- Operational importance
- Hydraulic capacity
- Required life of the pipeline
- Location of the main

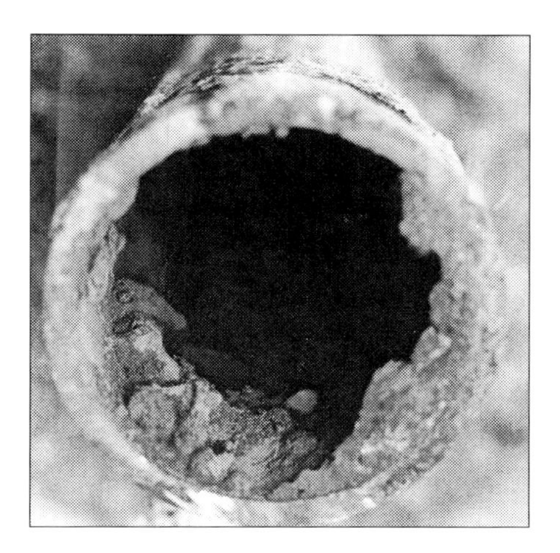

Figure 1 – Cast iron pipe

The UK water industry employs a range of rehabilitation techniques, selecting the appropriate option for each project. OFWAT report annually[1] on the split between mains relined and mains renewed, this offers an indication of industry trends over time. Figures are published by OFWAT for the period 1990-91 to 1997-98.

The global split in the UK between reline and renew is currently running at, 41% reline versus 59% renewal. This figure does not however accurately portray the split for individual water companies. An OFWAT report[1], shows that water companies have used policies ranging from, 0% reline and 100% renew, to 96% reline and 4% renew, over the 7 year period to 1998.

Potable water, trunk and distribution pipe systems have historically been constructed from a wide range of pipe materials. Figures 2 and 3 have been included to demonstrate both the ranges of pipe materials in use and the mains diameters employed in typical UK water company distribution networks. It might be interpreted from these figures that in the UK 75% to 80% of buried water company assets fall within the diameter range 50mm to 150mm and would account for some 250,000 km of pipe.

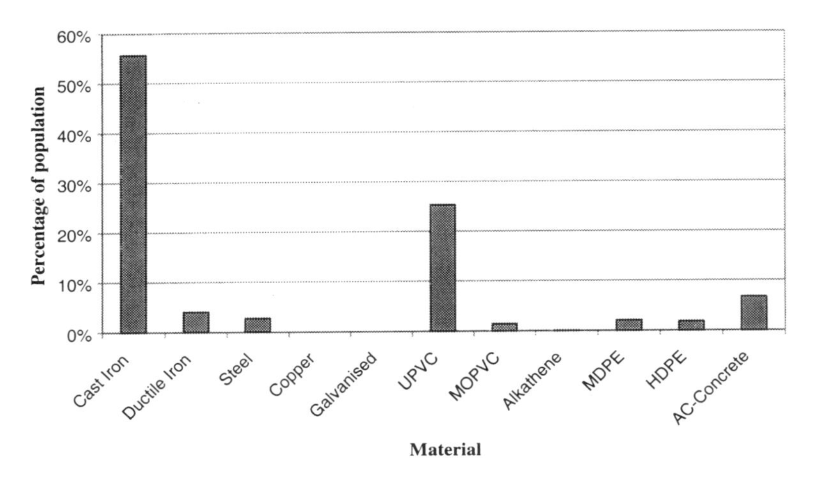

Figure 2 – Mains population by material

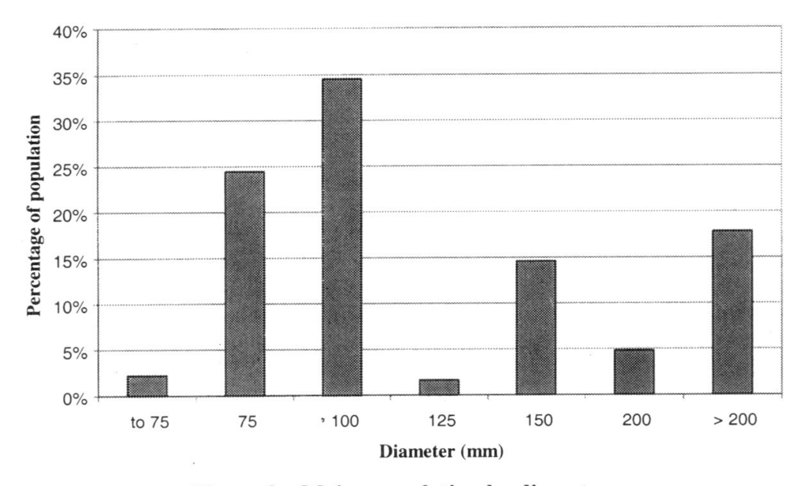

Figure 3 – Mains population by diameter

5 CONDITION ASSESSMENT TECHNIQUES

There is very little known about the detailed structural condition of water distribution networks in the United Kingdom. Water companies are now under increasing pressure to improve the management of their assets. It is important that future maintenance of networks is based upon an assessment of asset condition and associated risk of failure.

As condition assessment tools and techniques become more advanced they should help the water companies to ensure that investment in network maintenance is properly targeted. Where the consequences of pipe failure are potentially very serious in the gas and petrochemicals industries much work has already taken place to develop condition assessment, inspection and risk analysis techniques.

The assessment of pipe condition is generally based upon information gained from a number of direct and indirect sources using a variety of techniques. Water quality monitoring for example, would be considered as an indirect method, whereas in-situ pit depth measurement, is a direct method.

5.1 Sample assessment

The assessment of pipe samples is common practice by United Kingdom water companies. This involves removal of a pipe section for assessment and its substitution by a replacement length. The assessment process typically involves:

- Examination of external and internal condition and measurement of key dimensions
- Cutting of pipe longitudinally and photographing, Figure 4
- Grit blasting of pipe, internally and externally to remove all corrosion products
- Photographing internal and external surfaces after gritblasting
- Measurement of corrosion pit depths and extent of corrosion
- Testing of soil sample to assess corrosivity adjacent to the pipe
- Metallographical examination of wedge cut from pipe wall

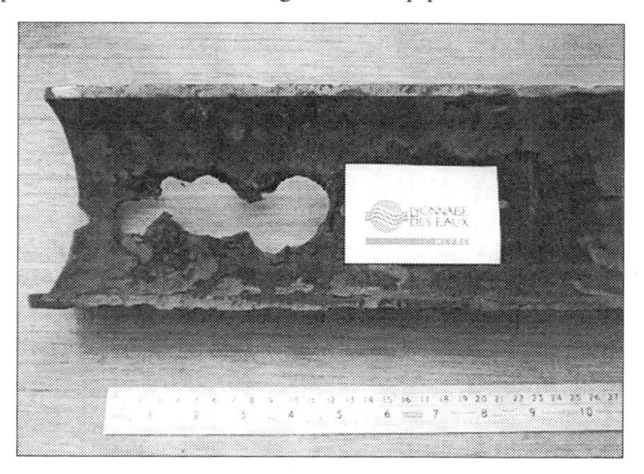

Figure 4 – Pipe sample assessment

5.2 Closed circuit television (CCTV)

Equipment for internal inspection of pipes using CCTV has been commercially available since the early 1970s. Today, CCTV systems are available that cover a broad range of pipe sizes from 25mm to 2,000mm diameter. In larger pipe diameters, inspection ranges up to 400m are now possible. Camera technology has also become increasingly sophisticated and many systems now offer colour, zoom, pan and rotate facilities.

As trenchless pipe rehabilitation technology has increased in popularity, CCTV has been used prior to rehabilitation as a tool to determine the presence of bends, changes in diameter, service connections and other obstacles. During the Epoxy Resin Lining of old Cast Iron water mains CCTV is used to identify potential obstructions, inspect the internal pipe wall after cleaning and subsequently to inspect the cured epoxy after lining.

5.3 Metal loss assessment

The first pigs to inspect pipelines for metal loss from the inside were developed for steel pipelines in the 1960s. · British Gas developed its first Magnetic Flux Leakage (MFL) inspection tool in the late 1970s[4]. The use of these pigs has historically been confined to the high-pressure oil and gas industry where the cost of inspection can be offset by the high product value. There are different requirements in the water industry where product value is much lower, distribution systems are complex in design and pipe materials are predominantly Cast Iron.

There are a number of conventional, well proven Non-destructive-testing (NDT) techniques that have been applied to the inspection of buried pipes. Those deemed to be most appropriate for detecting metal loss in buried pipelines, are covered by the following descriptions:

5.3.1 Magnetic flux leakage technique

When applied to the internal inspection of pipelines, MFL works by magnetising the pipe wall by use of permanent or electromagnets. A homogeneous ferromagnetic material will show a continuous magnetic field within the material. In the presence of a corrosion defect for example, the magnetic field is distorted, and this causes local magnetic flux leakage fields. The local flux leakage fields can then be detected using sensors.

5.3.2 Remote field eddy current technique

Eddy current internal inspection of buried pipes requires a coil carrying an alternating current (A.C.) to be passed over the pipe surface. By measuring changes in the coil impedance it is possible to identify the presence of flaws. The Remote Field technique[5] involves combing the eddy current exciter coil with a detector coil placed at least 1.8 pipe diameters away. This makes it possible to sense flux from the currents in the pipe wall. The magnitude and phase of the sensed voltage is dependent upon; pipe wall thickness, material magnetic permeability and the presence of discontinuities.

5.3.3 Ultrasonic technique

Using a probe the ultrasonic inspection technique transmits mechanical waves of short wavelength and high frequency into the pipe wall. A receiver monitors the transmitted wave and any attenuation caused by cracks, cavities and other discontinuities.

5.4 International water industry developments

As early as 1992 the American Water Works Association (AWWA) completed a study[6] into NDT of water mains. The study identified the Hydroscope tool from Canada as having the greatest potential; it is specifically designed for performing the internal condition assessment in grey cast iron and ductile iron water mains. The Hydroscope tool is reported[7] to be capable of detecting external and internal defects with pipe wall loss greater than 20% and defect diameter greater than 50mm. The tool is available in sizes suitable for use in pipe 150 mm to 400 mm diameter. Other inspection tools are now under development by the following companies; Lyonnaise des Eaux, BG Technology and Advanced Engineering Solutions Limited.

6 CONCEPT DEVELOPMENT

The assessment of pipe condition by removal and evaluation of pipe samples is by definition, not representative of general pipe condition, and provides the operator with only a very localised indicator. It is the author's view that water companies now require a simple, reliable and low cost, condition assessment method that will improve upon their current pipe sampling programmes. It is envisaged that this would allow for both routine and opportunistic assessment of pipe condition.

The direct measurement of pipe condition has a principle advantage over conventional sample assessment in that, it will be possible to detect defects that would otherwise be missed by not including them within a particular pipe sample length. The structural condition of pipes can vary over short distances, and a sample removed from a section of buried pipe for assessment may or may not include those defects that would for structural reasons make the difference between the choice to renew rather than reline.

When evaluating rehabilitation options with current assessment tools, water companies are limited by a decision to either renew or reline over an entire pipeline section. If the operator were to have a detailed assessment of pipe condition over the entire length of the same pipeline section then further options could be made available. The direct measurement of pipe condition using an in-pipe system could, for example, be used to identify the proportion of a pipeline section requiring non-structural rehabilitation. The remaining lengths showing evidence of a lack of structural integrity could then be treated as 'pinch points' and be subjected to localised structural rehabilitation solutions.

The United Kingdom population of older Grey Cast Iron mains exhibits the majority of burst failures, high leakage levels and water quality problems. It is these mains that are targeted for rehabilitation, inspection of them would allow for selection of the appropriate rehabilitation techniques. On a purely financial basis it could be difficult to justify such operations, the economic feasibility of iron mains inspection will depend on a number of factors, including:

- Mains diameter
- Location
- Can the main be taken out of service
- Criticality of supply through the main
- Risk and consequences of failure
- Maintaining serviceability to customers

All of the above contribute to the potential savings that can be achieved through the deferment of rehabilitation, and where required, the selection of appropriate rehabilitation techniques. Inspection of mains should aim for better targeting of the available funds, thus providing the maximum benefit and improvements in service to customers.

It is therefore recommended that any whole life costing approach to the selection of rehabilitation techniques should include an economic study into the options offered by these pipeline inspection technologies.

7 REFERENCES

[1] 1997-98 Report on the financial performance and capital investment of the water companies in England and Wales, OFWAT, 1998.
[2] 1997-98 Report on leakage and water efficiency, OFWAT, 1998.
[3] Water Industry Act 1991 - HMSO
[4] E Holden, 1997. The changing role of inspection. Pipeline Pigging conference. Amsterdam.
[5] TR Schmidt, 1989. History of the remote field eddy-current technique. Materials Evaluation, 47 (1): 16-22.
[6] American Water Works Association, Technology review non-destructive testing of water mains for physical integrity, AWWA, 1992
[7] D Russell, Meeting with hydroscope, 1999.

C571/038/99

The implications of an ageing asset base on water industry cost of capital

H BELL
Advanced Engineering Solutions Limited, Cramlington, UK

SYNOPSIS

The 1999 Periodic Review has generated a great deal of debate about what is an acceptable figure for the UK water industry's cost of capital. Essentially, it is determined by the ability of a company to meet its financial obligations, which in turn is influenced by risks facing a company, and the effect this could have on its cash flow.

The water industry is characterised by a high level of sunk costs in buried infrastructure assets. The maintenance of these assets, both sewerage and distribution systems requires large amounts of capital investment on a long-term basis. Any uncertainty as to the cost of maintaining the asset base increases the risk faced by the industry, and thus theoretically, will lead to an increase in the cost of capital.

This paper explores the cost of capital issue in the context of the water industry, focussing on the level of funding required to maintain an ageing asset base. It includes a summary of the Periodic Review process, looking at the regulatory significance of achieving a reliable estimate for the UK water industry's cost of capital.

1. INTRODUCTION

The 1999 Periodic Review has generated a great deal of debate about what is an acceptable figure for the UK water industry's cost of capital. If too low an estimate is used, water companies profits will be adversely affected, as will shareholders returns. Too high an estimate, then customers will lose out through paying higher bills than necessary.

Many factors will influence a water company's cost of capital. Essentially, it is determined by the ability of a company to meet its financial obligations, which in turn is influenced by risks facing a company, and the effect this could have on its cash flow.

The water industry is characterised by a high level of sunk costs in buried infrastructure assets. The maintenance of these assets, both sewerage and distribution systems, is essential if the industry is to satisfy the regulatory requirements and prevent problems arising in the future. This requires large amounts of capital investment on a long-term basis. Any uncertainty as to the cost of maintaining the asset base, increases the risk faced by the industry, and thus theoretically, could lead to an increase in the cost of capital.

A number of questions need to be answered such as:

- Can this risk be quantified?
- What affect is it likely to have on water industry cost of capital?
- What steps can be taken to minimise this risk?

This paper explores the cost of capital issue in the context of the water industry, focussing on the level of funding required to maintain an ageing asset base. It includes a summary of the Periodic Review process, looking at the regulatory significance of achieving a reliable estimate for the UK water companies cost of capital.

2. COST OF CAPITAL

2.1 What is it?

The cost of capital can be defined as the minimum return that providers of capital require to persuade them to invest in, or lend funds to, a business. Thus it has two components:

1. Cost of Equity – based on the return on investment required by shareholders.
2. Cost of Debt – based on the interest required by loan institutions on debt.

The weighted average cost of capital (WACC), is defined as the average cost of a company's loan and equity capital, weighted according to the proportion of debt and equity (i.e. gearing) that finances the company.

2.2 How is it determined?

The component cost of debt (K_d) is typically determined by considering the after-tax cost of new debt (K_{dn}), which is itself based on current and prospective future interest rates over the period of the loan. Thus, the cost of debt is given as:

$$\mathbf{K_d = K_{dn} (1 - T)}$$

(where T is the corporate tax rate)

One problem inherent in this approach is that it ignores the historical, or embedded, cost of debt from previously raised funds. The average cost of all debt raised in the past and still outstanding, could be much higher or lower than the current cost of new debt in an environment of unstable interest rates. This is particularly relevant to the UK water companies and is discussed further in **Section 3.3**.

The component cost of equity is the rate of return investors require on their share holdings. It can be estimated using one of the following approaches (1):

- Capital Asset Pricing Model (CAPM)
- Discounted Cash Flows (DCF)
- Bond-yield-plus-risk-premium method
- Dividend Growth Model

The CAPM is the most commonly adopted approach, and it has been used by Ofwat and others to calculate the cost of equity for the water industry. It requires estimates of 'Beta' (β), which is the movement of a company's share price relative to the stock market, and the 'equity risk premium' ($K_m - K_{rf}$), which is the return required by investors over and above the risk-free rate of return (K_{rf}). The equity cost of capital (K_e) is then given as:

$$\mathbf{K_e = K_{rf} + (K_m - K_{rf})\,\beta}$$

(where, K_m = Market Return)

Most companies will try to optimise their capital structure by achieving an appropriate balance between debt and equity financing. This balance will be determined by several factors such as interest rates, taxation, industry type and perceived risk. If W_d and W_e are the target weightings of debt and equity respectively, then the weighted average cost of capital (WACC) is given as:

$$\mathbf{WACC = W_d K_d + W_e K_e}$$

2.3 Why is it important?

A reliable estimate of a company's cost of capital is critically important for two key reasons:

1. In order to maximise shareholder value, the costs of all inputs, including capital, must be minimised.

2. Capital investment decisions require an estimate of the cost of capital to prevent a company from running into financial difficulties at a later stage.

In the UK water industry, another crucial reason for estimating the cost of capital is that it is used by the regulator when setting price limits that will determine the rate of return that the water companies can make on their assets.

In order to borrow money, companies must meet certain financial criteria, especially interest cover. Any unexpected financial shocks, for example, a requirement to invest heavily in unplanned mains/sewer replacement programmes, could mean that in a particular year, a company earns less than its cost of capital.

The cost of capital is particularly important to the water industry because of its need to finance large investment programmes on a long-term basis. The following comments, obtained from the annual reports of three UK water companies illustrate the importance of capital expenditure in water industry operations:

"The company will strive to achieve yet further efficiencies in operating costs and asset management. However, depreciation and interest costs will rise significantly as a consequence of the high capital investment to maintain and improve services, which is expected to peak around £570 million in 1998/99 following £516 million in the year just ended."
(Chairman's Statement, Severn Trent Plc. Annual Report & Accounts 1998)

"In 1997/98, we invested a record £471m – some 17% higher than last year – on improving, expanding and replacing our water and wastewater assets, bringing our total investment to £3bn since privatisation."
(Chief Executive's Review, Thames Water Plc.Annual Report & Accounts 1998)

"Total regulated capital investment was £350.1 million in the year and it is expected that a similar level of investment in service improvements will be maintained in the current year. Net cash flow from operation activities was insufficient to fund this level of investment and borrowing within the regulated water services business increased accordingly."
(Chief Executive's Review, Yorkshire Water Plc. Annual Report & Accounts 1998)

3. PERIODIC PRICE REVIEW

3.1 Setting Price Limits

The Director General of Water Services (DGWS) is responsible for setting water company price limits, and comparative efficiency information is used to identify best practise, which provides an incentive for companies to become more efficient. Within the regulatory framework, prices are set in such a way that the benefits of a water company's increasing efficiency is passed on to customers while at the same time allowing the company to finance its activities and provide a reasonable return for shareholders. This is why the Regulator needs a value for the water industry cost of capital (WICOC). For example, it has been estimated (2) that a one-percentage point change in cost of capital equates to £8.50 on the average household bill. With approximately 20 million water customers in the UK this equates to industry revenue of £170 million.

Increases in prices are limited by the RPI ± K formula, where K is the amount by which a company can increase (or must decrease) its charges over and above the rate of inflation.

K is a company specific factor comprising the following components :

Po	-	Past outperformance
X	-	Future efficiency gains
Q	-	Quality standards
V	-	Enhancements to the security of supply
S	-	Enhanced service levels

3.2 Review Procedure

Ofwat's incentive based regulatory regime places emphasis on outputs, i.e. making sure that the water companies deliver a good level of service to consumers at a fair price. The Regulator uses information on water industry capital unit costs to make a judgement on the appropriate level of future capital expenditure to be assumed in price limit determinations.

The timetable for the ongoing 1999 review procedure is as follows:

Phase 1 – Framework Development (February 1997-October 1998)

- In October 1998, Ofwat published its consultation paper, Prospects for Prices. This is central to the review and sets out all the relevant strategic issues that will be considered in the price setting process for the period 2000-2005.

Phase 2 – Decisions and Determinations (October 1998-November 1999)

- Water companies submit their information returns for the price review in addition to their draft Strategic Business Plans (SBPs) and Monitoring Plans (AMP 3). The draft SBPs outline the companies views on price limits.

- The DGWS issued draft determinations on price limits in July 1999, and the final determinations will be issued in November 1999.

Phase 3 – Implementation of Price Limits (December 1999 onwards)

- In March 2000, the water companies will publish their monitoring plans for the next five years.

- In April 2000 the price limits will be implemented.

The draft determinations have proposed an average P-nought price cut of 14%, with little or no price rises for the next five years. Consultancy firm Cap Gemini (3) estimated that these proposals would reduce the industry's turnover by 17%, and that unless costs were reduced by 29%, profits would fall by 43%. **Figure 1** shows the range of P-nought price cuts proposed by the regulator, compared with those requested by the Water and Sewerage Companies.

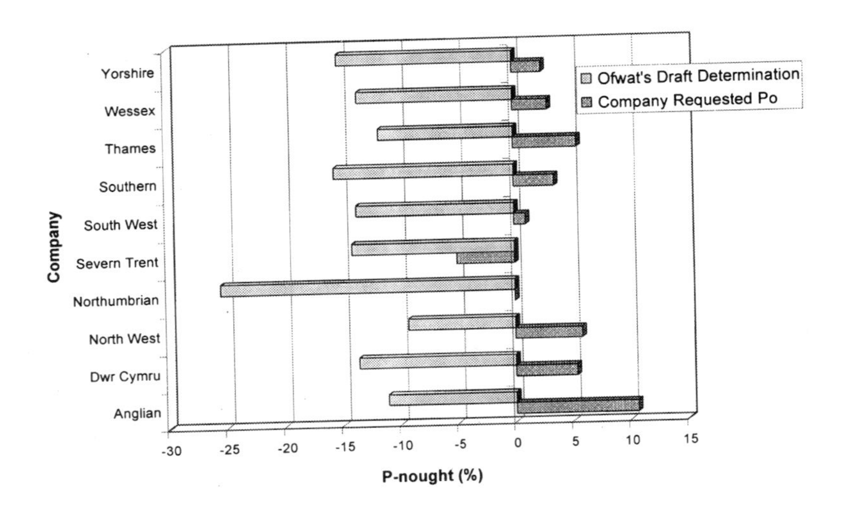

Figure 1 - Ofwat's Draft Determinations (P-nought)

3.3 WATER INDUSTRY COST OF CAPITAL DEBATE

In October 1998, Ofwat published its Prospects for Prices document (2), which among other things sets out its thinking on WICOC over the period 2000-2005. Despite the capital intensive nature of the water industry, the Director believes that it is perceived by investors as relatively low risk, and proposed a range of between 4.1% and 5.5% (see **Table 1**) for the post tax cost of capital for an efficiently financed water company.

Table 1 - Ofwat's WICOC assumptions

Component	Low Range Estimate	High Range Estimate
Risk free rate	2.5 %	3.0 %
Debt premium	1.0 %	1.75 %
Pre-tax cost of debt	3.5 %	4.75 %
Tax rate	30 %	30 %
Post tax cost of debt	2.5 %	3.3 %
Equity beta	1.2	1.5
Equity risk premium	2.75 %	3.75 %
Post tax cost of equity	5.8 %	8.6 %
Gearing	50 %	60 %
Post tax WICOC	**4.1 %**	**5.5 %**

One of the main criticisms is that the DG based his estimate of WICOC on current market data. However, others such as London Economics believe a longer term perspective is more appropriate. This is especially so when considering investment requirements in infrastructure assets. As noted in a recent press article (4):

"The water industry is a long term industry. Companies' decisions on investment and financing are based on long term considerations, and need therefore to be made in a long term context."

Following Ofwat's publication of Prospects for Prices, NERA conducted a survey of the financial markets to gauge opinions on a number of key issues surrounding the WICOC for the period 2000-2005. The respondents believed that Ofwat's approach was flawed on several counts.

Concerns were raised about the Regulator's decision to use a single cost of capital value for the industry as a whole. It has been argued that the cost of capital is likely to vary widely between companies due to a number of factors such as the assigned credit rating, management capabilities, size of capital expenditure programme and condition of the asset base.

Ofwat's initial estimate of the cost of debt did not take into account the water companies embedded debt cost. The general consensus from the NERA survey is that the cost of debt should be calculated as a weighted average of the embedded cost of existing debt and the expected cost of new debt. A representative from Natwest Capital Markets stated that:

"Water and sewerage companies invest in long term assets which it would be reasonable, even prudent, to finance with long dated debt. It is unrealistic to ignore the cost of historic debt particularly in a falling interest rate environment."

Ofwat has since modified its approach, and in the recently published draft price limit determinations, has made an allowance for the weighted industry average cost of embedded debt. Over the last decade, several water companies have committed to long term debt at what now appears to be high rates of interest. **Table 2**, which shows the outstanding eurobonds in the UK water sector which are due to expire beyond April 2000, illustrates the importance of the costs of embedded debt to water companies.

Table 2 - Outstanding Eurobonds in the UK water sector

Company	Bond Value	Start date	Term (years)	Nominal interest rate
Anglian	£100m	1990	23	12.0 %
Anglian	£150m	1996	10	8.25 %
Hyder	£125m	1996	20	9.5 %
Hyder	£150m	1996	10	8.75 %
Northumbrian	£200m	1996	10	9.25 %
Northumbrian	£200m	1997	10	8.625 %
Scottish Power	£200m	1997	20	8.375 %
Severn Trent	£125m	1991	10	11.625 %
South West	£150m	1992	20	10.625 %
Thames	£150m	1991	10	10.5 %
Welsh	£75m	1991	10	10.75 %

Source: "Debt Financing", Water Magazine No. 49 (16/04/99)

Ofwat's estimate of the cost of equity, based on the CAPM, was also considered to be low by several financial analysts surveyed by NERA. In particular, there was concern about Ofwat's use of historic beta values, which were not regarded as good indicators of current and future risks faced by the water industry. The financial community also considers Ofwat's estimate of the equity risk premium (2.75-3.75%) too low.

4. THE IMPLICATIONS OF AN AGEING ASSET BASE ON COST OF CAPITAL

4.1 Water Industry Asset Base

The water industry is characterised by a high level of sunk costs in infrastructure consisting of buried pipelines and associated assets. The networks were originally constructed in the Victorian era, and have been significantly added to over the years. Because of this, valuing the system on a historical cost (i.e. as laid) basis does not provide any meaningful figures. A more appropriate method to put a value on these assets is to estimate the cost of constructing an equivalent system at today's prices. This is known as the 'modern equivalent asset' MEA value. In August 1998, each of the water companies was required to provide Ofwat with an assessment of MEA value of its asset stock for the purposes of the 1999 Periodic Review. The bulk of the investment is tied up in water main networks, which for the industry as a whole have a gross MEA value of £37.8 billion.

For the purposes of the Regulator's Periodic Review, the water companies assets (excluding sewerage services) are divided into the following categories:

- Water Resources
- Water Treatment
- Storage
- Pumping Stations
- Water Mains
- Management and General Assets

For the average UK water company, the total MEA value may typically be allocated to each of the categories as shown in **Figure 2**.

The Regulatory Capital Value (RCV) is the value of the assets upon which the regulator allows a projected rate of return when setting prices. In the UK, the regulatory value of the water industry's assets is estimated to be 10% of the MEA value. The difference between these two figures is called the 'capital value discount', and it was reflected in the flotation price of the water companies which, at privatisation in 1989, were sold for a fraction of the replacement cost of their assets. Hence, the capital value discount is a permanent feature of the UK water industry, and has an indirect influence on the price limits set by the Regulator. The capital value discount effectively prevents potential new entrants from developing their own water distribution networks, since it would not be possible for a newcomer to both compete on price with incumbents, and earn a suitable rate of return on the investment required in new infrastructure.

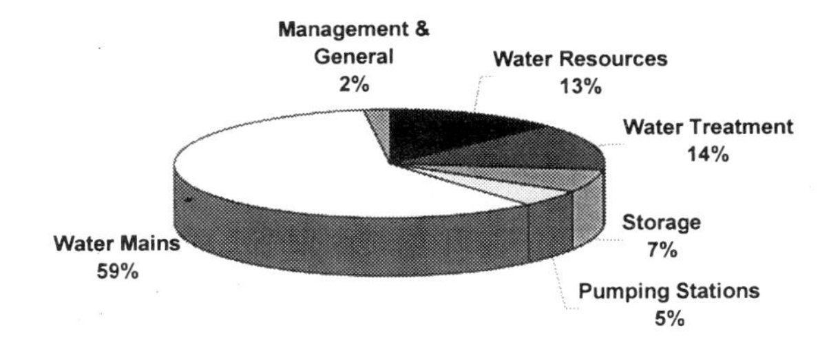

Figure 2 – MEA Value by Function

4.2 Risks Faced by the Water Companies

The water companies are faced with four types of risk:

- Cost risk – unit costs will differ from those assumed
- Demand risk – demand will differ from that assumed when setting price limits
- Performance risk – quality/service level performance will fail to meet standards
- External risk – risk factors beyond the control of the companies

In terms of maintaining an ageing asset base, cost risk and performance risk are the main areas of uncertainty for the water companies.

4.2.1 Cost Risk

When setting price limits, Ofwat uses a comparative capital unit cost approach known as the 'cost base' to identify the efficient/inefficient water companies. Companies with higher capital costs are considered to have more scope for savings in their expenditure projections than companies with lower capital unit costs. However, many in the water industry believe that Ofwat's method for determining unit costs is fundamentally flawed (5). An independent study carried out by consultants Ove Arup, showed that the specifications for unit costs supplied by Ofwat to the water companies are too vague, and that some items that could have a significant impact on capital spending (e.g. IT costs) are ignored altogether.

4.2.2 Performance Risk

Essentially, performance risk is that quality and service level performance will not match the regulatory requirements requiring additional capital expenditure. One of the major contributing factors to performance risk is that water companies are not able to predict with any certainty, the remaining life span of their ageing pipeline system, much of which was constructed in the Victorian era. Therefore, those companies with networks containing a high proportion of ageing pipelines are likely to be perceived as being more risky investments, than those with newer systems. This attitude is reflected in a recent statement issued by Standard and Poor as they issued a single-A-minus (i.e. relatively low-risk) long term rating to a recent £350m bond issue by Anglian Water:

"Anglian operates a cost-efficient and relatively new distribution asset base. The good condition of the system is reflected in the company's high level of compliance with regulatory standards."

Some industry observers also believe that the way the water companies are regulated in the UK places too much emphasis on short-term rehabilitation techniques (e.g. S19 water quality programmes) which is storing up problems for the future. Consider, for example, the fact that the current rate of replacement for some companies indicates an assumed life of more than 1,000 years for distribution pipes and 4,000 years for sewers (6). Clearly, this level of replacement is unsustainable, and at some point in the future significant levels of investment will be required to ensure that the required levels of service are maintained.

If a company finds that it has to spend a significant proportion of its turnover on unplanned maintenance (due to unexpected service failures) to replace underground assets, then its cost of capital is likely to rise. This is because:

1. The company will need to increase its debt to cover the costs of increased maintenance, and a greater proportion of the company's cashflow will be required to service the increased debt. As a consequence of this there will be less money available to distribute to the shareholders as dividends.

2. As cash flows become more unpredictable the company will be perceived to be a higher risk investment, causing its cost of capital to rise.

5. RECOMMENDATIONS AND CONCLUSIONS

The majority of the older assets in the UK consist of networks of cast iron pipelines, and corrosion is the main factor leading to their failure. However, the rate of corrosion in cast iron pipes is a random process, dependant on several factors, and therefore it is extremely difficult to predict the life of these assets with any degree of confidence. In order to assess current and future asset maintenance requirements, the water companies need to acquire a detailed knowledge of the condition of their buried infrastructure. This would enable the risk of failure to be predicted with more accuracy.

In the gas and petrochemicals industries, where the consequences of failure are much higher, a lot of research has been carried out to develop pipeline inspection techniques for condition assessment purposes. As condition assessment tools and techniques for the water industry become more advanced, it will be possible for the UK water companies to establish their investment requirements with a higher level of confidence, and take steps to ensure that investment in network maintenance is properly targeted. This will reduce the operational uncertainties faced by the industry and should bring down the cost of capital.

REFERENCES

1. Financial Management Theory and Practice (8th Edition), E.F. Brigham & L.C. Gapenski, 1997

2. Prospects for Prices – A consultation paper on strategic issues affecting future water bills Ofwat, October 1998

3. "Price Review to force huge cuts in costs", Utility Week, 30th July 1999

4. "Paying the price of debt", Richard Hearn, Water No. 44, 5th March 1999.

5. "Infrastructure specs flawed", Water No. 35, 18th December 1999

6. A Price Worth Paying – The Environment Agency's proposals for the National Environment Programme for Water Companies 2000-2005, Environment Agency, May 1998

C571/039/99

Structural assessment of iron mains

D SMART and **P WARD**
Inspection Services Department, Advanced Engineering Solutions Limited, Cramlington, UK

ABSTRACT

The economic regulation of expenditure on maintaining and improving the low pressure Gas and Water pipeline networks in the UK is focusing the utilities on developing cost effective strategies for the rehabilitation of these valuable assets.

This paper briefly considers the design of low pressure iron pipe systems and the significance of their structural deterioration in terms of the range of loading regimes they may undergo. An approach to structural condition assessment currently employed by Advanced Engineering Solutions Limited is outlined. This approach identifies the most critically loaded sections of a pipeline and uses the most technically and economically appropriate NDT techniques to determine the condition of these sections. A statistical analysis based on the critical loading conditions for the pipeline is then used to provide a sound estimate of its performance capability under the applied loading regimes.

1 INTRODUCTION

There are hundreds of thousands of kilometres of low pressure pipelines in service within water and gas distribution networks around the world. These utility pipe systems often date from the mid 1800's onwards and they therefore include a high percentage of pipelines constructed using poorer quality materials. For example, pit or spun cast grey iron pipe which is a fairly brittle material due to its flaked graphite structure, has been in common use over the majority of this time. Where more modern materials have been used, for example steel and ductile iron, the latter having a spheroidal graphite structure giving it superior tensile strength and ductility to grey iron, there has been variation in the quality of the construction and maintenance techniques applied. This can lead to specific deterioration problems.

With the introduction of competition, common carriage, and more fiscal influences in the management of these pipelines, the approach to their maintenance has by necessity changed. It is necessary for the pipeline owners to understand the structural condition of their pipe systems, particularly those sections of the network which are operationally critical. This allows the proper planning of repairs and replacement activities.

This paper briefly considers the design of low pressure metallic pipe systems and the mechanisms causing their structural deterioration. It outlines the range of loading regimes they may undergo and discusses techniques for assessing their structural condition.

2 LOW PRESSURE PIPE SYSTEMS

Low pressure pipe systems operated by the utilities range in diameter from some 3" up to 48" or more. By example, the total water main network asset stock for England and Wales is shown in Table 1 below [1].

Table 1 Length of Water Mains (Km) by Diameter (mm)

less than 300	300 - less than 600	600 - less than 900	greater than 900	Total
246,000	51,000	20,000	8,000	325,000

Low pressure pipe systems differ in a number of ways from the higher value, high pressure pipelines, particularly in their use of lower quality materials and mechanical jointing, and they often have little, if any, active maintenance programmes.

A major point to recognise when considering the performance of low pressure pipelines is the similarity in design approach taken, regardless of the fluid they are carrying. Since these pipelines are designed to accommodate relatively low operating pressures, the external loading on buried pipelines can, in terms of failure, be as significant as the internal pressure loading. It is therefore essential that these pipelines are designed to carry significant external loading as well as the pressure loading, and any structural assessment applied must take account of both these loading mechanisms.

Failure records are often a major driver in the selection of pipelines for maintenance. The failures occurring are generally very dependent on the beam strength of the pipes, which is related to their diameter and wall thickness. Small diameter iron pipes have low beam strength and a poor failure record. Large diameter pipes often have a good failure record since they have relatively high beam strength and are less sensitive to external loading.

Figure 1 indicates failure levels for different diameters of iron pipes. Data was obtained for two areas within a UK water distribution network. The figure clearly illustrates the relationship between failure rate and pipe diameter.

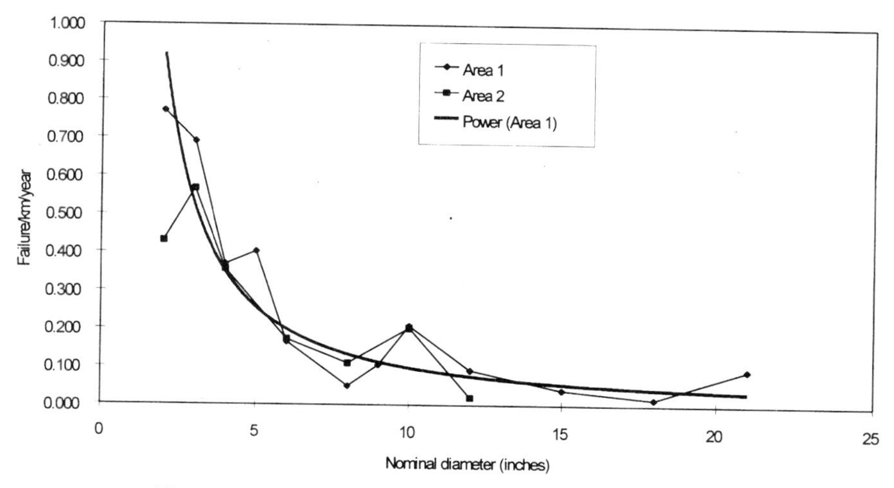

Figure 1 Cast Iron Failure Rates/Nominal Pipe Diameter

The approach taken to the maintenance of low pressure pipelines should therefore be influenced by their diameter as well as their criticality, as discussed below.

3 PIPELINE MAINTENANCE

The fundamental factors influencing both the gas and water industries when considering maintenance or replacement are the extent, size and age of the pipe systems, the range of materials involved, and the criticality of the pipeline to the continued provision of service to customers.

Older systems covering a large area are often characterised by:-

- a fairly high number of in-service failures being, by necessity, accepted operationally,
- management of repair and maintenance activities often being focussed more towards failure repair than failure prevention,
- replacement activities being a necessity, but being driven by strategic issues, due to the extent of the systems.

3.1 Strategic Replacement and Maintenance Strategies

A selective approach to the maintenance and replacement of these systems is generally taken. This selective approach is based on business and operational requirements, but it must also meet the economic requirements of the Regulator. Levels of service to the customer in the UK are now defined by the Regulatory authorities, who also exact a level of control over the capital expenditure. In the water industry in particular, this deliberate external influencing of the utility businesses is intended to introduce level of competition between what are effectively regional monopolies.

The strategic approach taken by the UK gas and water industries to system maintenance and replacement differs.

- The gas industry has adopted a relatively complex replacement policy which is based on removal of the pipe which is considered to be most at risk of failure and for which the consequences of failure present a potential safety hazard. In the UK, these policies are policed technically by the Health and Safety Executive, although the 'Energy Regulator' influences the total financial investment. The general approach has been to remove a great deal of small diameter iron pipe to minimise system failures.

- The UK water industry does not have the same safety drivers as the gas industry but has the same technical difficulties with its ageing pipe systems, as well as regulatory and customer service drivers. Over the past decade the industry's major objective has been to meet the water quality targets set by the Drinking Water Inspectorate. The major part of the funding to the industry has also been based on meeting the water quality requirements. This has involved cleaning and lining a great deal of the small diameter iron pipe to prevent the potable water being contaminated by the corroding iron.

Therefore, although the requirement to keep a large stock of old pipe system in service is the same for both industries, the approach has differed. However, when the diameters of these pipe systems are taken into consideration, it can be seen that the failure statistics and the influence of diameter on these has influenced both industries. In both cases the great majority of ongoing work has been to replace or maintain the small diameter pipes, whilst the larger diameter pipe, above around 12" diameter have had limited on-going maintenance.

Operationally it is important to recognise the criticality of large diameter pipes, since they normally supply major sections of the network and hence large numbers of customers. They have served well, although their 'operational criticality' is now being increasingly recognised by utilities. There is increasingly a need to understand their structural condition and hence probable future performance. This allows their current financial value to be assessed and the expenditure necessary to extend their life to be estimated.

3.2 Operationally Critical Pipelines

When considering the specific problem of assessing and ultimately maintaining larger diameter operationally critical pipelines, it is obvious that their strategic replacement will generally be economically and strategically very difficult if not impossible. However, the cost of in-service failures can be very high and therefore there is an increasing need to understand the structural condition of these pipes. Understanding their structural condition can allow informed operational decisions to be made, for example:-

- focused replacement of sections of these pipes,

- local strengthening or repair to eliminate predicted failures.

When considering how best to assess the structural condition of pipelines, it is necessary to understand how they deteriorate, and how this deterioration can lead to failure.

4 DETERIORATION OF IRON PIPELINES

The major deterioration mechanism found in iron pipelines is corrosion. The types of corrosion encountered on iron pipe are shown in Table 2 and summarised below.

Table 2 Types of Corrosion Occurring in Cast Iron Pipe	
Pit cast iron (grey)	**Spun cast iron (grey or ductile)**
General corrosion	General corrosion
Pitting corrosion	Pitting corrosion
	Fissure corrosion

4.1 General Corrosion

General corrosion can occur on the internal and external surface of water pipe, and is produced by a surface chemical reaction, resulting from the presence of moisture and oxygen or chemicals. This results in a localised reduction in the wall thickness. Iron pipe can suffer a rate of radial metal loss in the order of 0.1 mm to 0.2 mm per annum. However, higher rates of general corrosion can occur under certain environmental conditions.

4.2 Pit Corrosion

In some circumstances, for example where a pipe coating is damaged, the corrosion attack can be very localised forming a deep pit in the pipe wall. Experience indicates that large pit corrosion occurs more often in ductile iron pipe than grey iron, which displays a more uniform level of general corrosion.

4.3 Fissure Corrosion

Subjecting a spun grey iron pipe to tensile stresses in the presence of corrosion can result in a particular type of structural weakening or failure, known as fissure corrosion. The onset of fissuring occurs when the tensile stress level in the pipe exceeds 30% of its ultimate tensile strength. Fissure cracks appear circumferentially in the smaller diameter pipes and longitudinally in the larger diameter. This can be expected due to the applied loading regimes, with bending producing longitudinal tensile stresses, and crushing producing tensile stresses in the hoop direction. The importance of these cracks is that they reduce even further the structural capability of the pipe wall, increasing the potential for sudden failure.

Fissure corrosion can occur in both ductile and grey iron pipe which has been manufactured using the spun casting process. This failure mechanism results from the combined action of stress and corrosion producing fissures that travel through the pipe wall, the rate of growth being related to the stress level in the pipe wall. Fissures are not formed in the absence of a tensile stress.

It is known that grey iron pipes can be penetrated by fissures at a higher rate than general corrosion would occur, inducing a marked reduction in the resistance of the pipe to any applied bending stress. Whilst ductile iron suffers fissure corrosion, its better strength and ductility properties make it much more resistant to fissure growth than grey iron.

In practice, where fissuring already exists in a pipe, reducing the wall thickness will increase the stress and increase the rate of fissuring.

5 ASSESSMENT OF STRUCTURAL CONDITION

Large diameter pipelines may have been constructed using one of a range of materials, although the most common and problematic materials are grey iron, ductile iron and steel. The requirement is to assess their structural condition when little or no access to the pipeline is available, since it is a normal requirement for them to remain in service.

A number of techniques are currently available for condition assessment of metallic pipes and these are described in detail in WRc and CIRA publications[(2),(3)]. A range of these techniques are discussed below.

5.1 Pipe Section Removal

The conventional approach used by many water pipeline operators is to remove and examine a small number of representative sections of pipe. This examination can include dimensional measurement, grit blasting to remove corrosion products and measurement of the remaining metal to identify remaining structural strength. It is common practice to take a small number of samples and to infer a great deal from these in terms of the overall pipe condition. Removing a statistically significant number of samples from a pipeline can be both technically unacceptable and economically prohibitive.

Determining the number of samples required to provide a statistically significant representation of pipeline condition is a much debated subject. Statisticians will always prefer more data, whilst in practice, the number of samples that can be provided will be limited by operational and economic constraints. Additionally, it is not always possible to determine whether the number of samples taken is statistically significant without information on the variation of pipeline condition along its length i.e. some form of comparator.

5.2 External NDT

Increasingly, non-destructive testing (NDT) is being used as an economic tool in the predictive maintenance approach to pipeline assessment, allowing extension of pipeline life based on planned replacement or repair. This approach has been applied to offshore pipelines where the asset value is high and the application of the technology can be economically justified. For lower value iron pipe systems, further development of the technology will be required to simplify the equipment and reduce the costs of usage.

The application of NDT for integrity assessment of steel pipelines normally requires full wall coverage in order to detect and measure critical defects, and the majority of existing NDT techniques are designed to achieve this. However, for iron pipelines where internal access may not be achievable or where external access is restricted, a more limited survey is normally accepted. A variety of new and improving NDT screening methods are entering the market which are suitable for indicating areas requiring further NDT evaluation. The following techniques could be considered for in-service inspection of sections of pipelines. Most of these techniques would be suitable for risk based inspection techniques.

5.2.1 Radiography

Pipe inspection crawlers using both x-ray and gamma ray sources are commercially available for pipe diameters of 6" and above. This technology is well proven and is in common usage for inspection of welds on steel pipelines both on-shore and offshore.

A low energy real time radiography technique has been developed in which the x-ray beam tangentially illuminates the wall of the pipe from the outside. Radiation is detected by an image intensifier upon which the metal loss or corrosion produces a shadow. This method can be used to detect corrosion but it can not quantify it.

A high energy real time radiography system does exist for full inspection of liquid filled pipes less than 12" diameter. The high energy beam is emitted from one side of the pipe and detected on the opposite side of the pipe. Images are electronically stored and can be processed to enhance corrosion and quantify pit depth and remaining wall thickness. Such a system needs a robot for remote application, due to the high radiation levels it emits.

Modern equipment is designed to be very safe, to minimise the risk from radiation. However, the equipment is expensive and operators must be suitably qualified. The use of this equipment in urban environments would pose some difficulties.

5.2.2 Pulsed Eddy Current

A non contact, low frequency, deep penetration eddy current system has been developed for external inspection of pipes with external lagging or external encrustation. The system has been developed and patented by ARCO, USA and their system is known as TEMP [4]. The method is reported to have very good reproducibility and accuracy but it does not detect very localised corrosion as it averages the information over an area of a few square inches. The system is not dynamic and requires several seconds for each measurement.

5.2.3 Ultrasonic Measurement

Ultrasonic systems are traditionally used for accurate and direct measurement of metal loss on steel pipelines. The technology can be applied using internal inspection pigs or external spot measurement equipment. For external measurement of buried iron pipes, there are two limitations to the use of standard ultrasonic probes.

- The metallic structure of iron is ultrasonically more noisy than steel and requires the use of a lower frequency of operation to obtain useful reflection signals from the pipe wall. Typically, the frequency will need to be reduced from 5 MHz to 1 MHz. Whilst a useful signal can be produce, the reduction in frequency results in a corresponding reduction in measurement accuracy and resolution.

- Ultrasonic probes require a good contact with a smooth pipe surface to provide an accurate measurement. The condition of the external pipe surface on buried pipes is usually very poor and therefore a significant amount of time and effort is required to prepare the surface prior to use of an ultrasonic probe.

5.2.4 Long Range Ultrasonic System

The long-range ultrasonic system has been developed for inspection of pipe sections with limited access. The technique uses bulk wave transducers which can inject an ultrasonic pulse into the pipe wall from the outside and the bulk wave travels longitudinally along the wall producing reflections at points of internal or external corrosion of the pipe wall. The system is capable of inspecting the wall over a distance of 1m from a single position. The technique is not capable of depth sizing or discrimination between internal and external corrosion. This technique is not yet matured and is still in the validation phase.

5.2.5 Ultra Long Range Ultrasonics

Recently, ultra-long-range ultrasonic systems have been developed. These use lamb waves to detect anomalies up to 50m from the measurement point. At least two systems are nearing commercial application, however, performance has still to be assessed.

5.2.6 Magnetic Flux Leakage

The use of magnetic flux leakage for inspection of steel pipelines is a well known and mature technology and is extensively used in inspection pigs for internal inspection of pipelines. This technology can be equally well applied using an external inspection tool for those pipes where appropriate access is available and the use of an internal pig is not possible. Advanced

Engineering Solutions Limited (AESL) are currently developing an external MFL inspection tool for use on iron and steel pipes.

5.3 Inspection Pigging

Although pigging technology is commonly applied to the inspection of gas and oil pipelines there is a definite reluctance to apply this technology to potable water pipelines because of economic, water quality and technical issues.

- The economic issues are related to the apparent low value of these water pipelines when compared to the high value of high-pressure gas and oil pipe systems. This means that unless the inspection technology can be manufactured and applied at low cost, then replacement or lining may be seen as a more economic option than inspection and the subsequent selective repair.

- Water quality issues are related to the potential disturbance to pipe wall deposits carried by the pig passing along the pipe. Overcoming water quality problems caused by operating pigs in pipelines is both a technical and an operational problem.

- Launch and retrieval of the pig system is generally difficult and expensive in the urban environment.

- Pig control and pig dynamics can be a perceived problem area, in an industry with little experience of pig operations.

Traditionally, only steel pipes operating at high pressures and carrying high value fluids have undergone regular condition assessment surveys. Prior to around 1975 it was necessary to validate the structural integrity of these pipelines using a hydrostatic test. During the early 1970's PIG mounted NDT techniques were developed to allow the pipeline to remain in-service during periodic validation following the initial pressure test validation. This is now the accepted approach to validating high-pressure, safety critical pipelines, although developments continue in both validation strategies and inspection technology.

The factors which make this approach acceptable on steel pipelines are:-

- the steel used in their manufacture is a homogenous material of high quality,

- the steel is primarily stressed to defined levels by internal pressure loading, and other loadings, such as overburden, are far less significant than for low pressure pipelines,

- external pipe loading can generally be considered negligible when compared to the internal pressure loading,

- failure mechanisms are well understood, allowing the inspection systems to be designed and calibrated to identify defects of critical size,

- the pipelines can be protected using high quality coatings and applied CP systems, minimising any on-going corrosion and simplifying the targeting of inspection technology.

When comparing the materials and techniques used for the construction of low pressure gas and water mains with those for these high pressure steel pipelines, significant differences are recognised:-

- the age of the pipe systems may mean that modern construction standards are not relevant,
- active corrosion control is not applied in the older pipes resulting in extensive and unpredictable corrosion levels,
- the iron materials are less homogenous and the criticality of the defects is more difficult to establish.

6 PIPELINE LOADING REGIMES

It should also be recognised that whilst the pipe condition can vary along the length of a pipeline, the loading regimes can also vary significantly along the length of low pressure pipelines. Therefore, when considering the structural condition of these pipelines, or more accurately their operational capability, a number of points related to the loading conditions must be considered.

- What combination of loading is acting on the pipeline along its length?
- What is the pipe wall condition along the full length of the pipe?
- Are there any points where the combination of applied loading and reduced strength pipe will cause failure?

The major loading regimes on water pipe systems are summarised in Table 3.

Table 3 Loading Regimes on Potable Water Pipelines

Loading Regime	Source	Comments
Internal (pressure) loading	Pumped pressure Head pressure Surge pressure	Controllable Possibly controllable Controllable
External (ground) loading	Traffic Ground movement Burial depth (overburden) Thermal and moisture induced Adjacent civil engineering works	Estimated Estimated Fixed Estimated Estimated and possibly controllable
Mechanical supports	Spacing of structural supports Movement of structural supports	Estimated and possibly controllable

The range of loading regimes above do not act on all pipelines and their magnitude will vary along the length of the pipeline. The approach taken to condition assessment must therefore recognise this potential for significant variations in both pipe condition and pipe loading along its length.

7 AESL APPROACH TO STRUCTURAL CONDITION ASSESSMENT

For critical low pressure pipelines, it is worthwhile considering an alternative approach to assessing their structural condition. The approach taken by AESL is to:-

- use a modern structural analysis technique to consider the applied loading along the entire length of the pipeline, from both internal pressure and external ground loading,
- model the pipeline with an assumed value of pipeline corrosion, based on a knowledge of the corrosion mechanisms and expected corrosion rates,
- identify those sections of the pipeline which are more critically loaded,
- assess the actual pipe wall corrosion levels in these critical sections,
- determine the criticality of the loading and pipe wall condition on these critical sections.

It is important when modelling to ensure that any additional loading applied to the pipeline due to changes in surrounding structures over its lifetime, are recognised and addressed. It is also important to understand and model the range of potential failure mechanisms and the factors which influence these.

When using this approach the fundamental requirement is to determine:-

- the pipe wall condition with some level of accuracy,
- all the applied loading regimes in any critical sections,

Both these requirements can be met based on a combination of theoretical values, measured values and estimates based on previous experience. The pipeline loading can be identified with some level of confidence, although the variability in applied loading due to ground conditions should be recognised. Pipe wall condition must be assessed with minimum intrusion into the pipeline. That is the value of the small amount of information available from local NDT or sampling must be used strategically. This requires that the approach must be based on a statistical analysis, which provides both an indication of the structural condition of the pipeline, and of the probable accuracy of this prediction.

When determining the condition of the pipe wall an approach is used which provides a choice of external inspection using NDT techniques, or pipe wall sampling. The approach used by AESL is to apply a statistical analysis not only to determine the sampling requirements but also to focus sampling towards those sections of pipeline where the loading has been identified as being more critical. In effect, AESL apply a version of the limited state design philosophy to these pipelines when addressing both the modelling and sampling criteria.

This approach can be applied to complete pipelines or to specific sections of pipeline which are recognised by the Water Company as being more operationally critical in terms of their supply parameters.

8 CONCLUSIONS

Condition assessment as used on high pressure pipelines is not directly applicable to low pressure pipelines, both technically and economically.

Cut out sampling may be operationally and economically impractical, if sufficient data is collected to be statistically representative of pipeline condition.

A combination of statistically relevant pipe wall assessment, based on an understanding of failure mechanisms and performing appropriate structural analysis is a more practical approach.

AESL use a combination of:-

- identifying the most critically loaded sections of pipe,
- using the most technically and economically appropriate NDT technique,
- applying a statistical analysis based on the specific critical defect regime for the pipeline.

This allows a sound estimate of the performance capability of these pipelines under the applied loading regimes to be made.

REFERENCES

(1) Serviceability of The Water and Sewer Networks in England and Wales, Information Note 35A, OFWAT, February 1999).

(2) Guidance Manual for the Structural Assessment of Trunk Mains, WRc, 1992.

(3) Water mains: guidance on assessment and inspection techniques, CIRA Report 162, 1996.

(4) MFL and PEC tools for plant inspection, J.H.J. Stalenhoef and J.A. de Raad BANT/KINT Biennale, Liege, Belgium, 23/24 April 1997, Lastijdschrift No. 53/2, July 1997.

Low Pressure:
Replacement Policy

C571/013/99

A mains replacement strategy – selecting the right mains for replacement

R K McALL
BG Technology, Loughborough, UK
D HAMLING
Transco, Gloucester, UK

Abstract

The UK gas distribution system operating below 7 bar pressure, consists of a network of approximately 250,000 kms of gas mains of various materials, mainly iron and polyethylene. Iron mains can fail through fracture, corrosion and/or joint leakage, and Transco has developed a programme of replacing iron mains, primarily with polyethylene, in order to maintain the integrity of the system, particularly with regard to safety.

Transco, as the pipeline operator for BG Plc, has to prioritise the replacement of individual mains in order to ensure that those which pose the greatest risk are replaced first and investment is targeted most effectively. Over the past 25 years, Transco has managed the prioritisation of mains replacement using a variety of approaches. The current scheme which has served the industry well in reducing incident levels is due for replacement with a more sophisticated tool.

By carrying out a statistical analysis of failure data from the distribution system collected over a period of 15 years, statistical models have been developed for Transco by BG Technology which can predict the likely probability of any given main failing and potentially causing an explosion incident.

Using the outputs from these models, Transco will be able to target those mains which pose the greatest risk for replacement, thereby maintaining its high safety standards in the most cost effective way.

1. Introduction

1.1 The mains network

Transco operates a distribution network that is over 250,000 kms long and delivers gas to around 20 million domestic, industrial and commercial gas users. The system operates at pressures up to 7 bar, but the majority of mains are in the low pressure range operating below 75 mbar. The mains population is comprised of the following materials:

Polyethylene (PE)	108,700 kms
Cast Iron	104,900 kms
Ductile Iron	21,400 kms
Steel	18,000 kms
Other materials	500 kms

Over half of the system is of metallic construction, the majority being over 50 years old, with some in excess of 100 years, (see Figure 1).

Pit cast is the oldest type of cast iron main in use. The wall thickness is relatively large as the pipes were cast vertically around a central core. In most cases the central core was not perfectly aligned and this caused the wall thickness to vary around the circumference of the pipe.

Spun cast is usually referred to as cast iron, this type of pipe replaced the pit cast type. They are more uniform in construction having a more constant wall thickness. Spun cast iron pipes were manufactured by casting the pipe in a spinning water cooled steel mould.

Ductile iron pipes use a different cast iron composition, which is less brittle than the older cast iron. The manufacturing process was similar to spun cast iron. The improved strength and ductility of the material meant that the pipe was produced with smaller wall thickness than the older cast iron pipe.

Steel pipe has a strong resistance to bending stress, although if not protected it is susceptible to corrosion.

1.2 Failure Modes

Fracture

Cast iron is a brittle material and if subjected to a bending stress, (for example as a result of ground loading) it will break. The pipe may be stressed as a result of sagging where the support for the pipe is reduced, or by hogging. In some cases the stress due to hogging may be increased and concentrated if the pipe base rests on a solid object such as a buried brick or rock. For smaller diameter mains (< 12 inch), the resulting fracture is usually circumferential. Even when new, cast iron pipe is subject to failure in this mode if the stress is sufficiently great. In most cases, cast iron pipe will have suffered some degree of corrosion and this will have the effect of decreasing wall thickness and so the pipe will fail under a smaller load than if in perfect condition. Large diameter (>12 inch diameter) cast iron mains are inherently stronger than smaller diameter and fractures occur less readily. The fracture pattern in large diameter pipe can be different with circumferential fracture becoming less likely, and longitudinal fracture occurring. This can lead to a part of the pipe wall breaking away.

Ductile iron is much more resistant to bending stress than cast iron and fractures (while still possible) are much less likely to occur.

Corrosion

Corrosion can occur on all buried iron pipe. The extent of this corrosion will depend on a number of factors, but is determined to a large extent by the corrosivity of the surrounding soil. The effects of corrosion on cast iron and ductile iron are different. In cast iron the corrosion products retain some residual strength, and, to outward appearance, the pipe may appear intact. However, in some cases, removal of the corrosion product shows that in reality there may be very little metal remaining. This renders the pipe more susceptible to fracture, or alternatively to the formation of a corrosion hole in the wall of the pipe when the pipe is disturbed.

The corrosion product on ductile iron has little residual strength which means that through wall corrosion leading to leakage is more likely to become apparent. In some cases substantial corrosion holes can form resulting in a significant leak. This is usually where the pressure differential between the inside of the main and outside is affected as a result of some disturbance such as ground loading, nearby excavation or even pressure variation within the pipe.

A particular form of corrosion which occurs in spun cast iron pipe (and to a limited extent in ductile iron pipe) is fissure corrosion. Fissure corrosion can occur when corrosion and significant stress levels are both present simultaneously. In this situation corrosion can take the form of fissures which extend much more deeply into the pipe wall than the general corrosion which may also be present. These fissures form across the pipe wall and substantially reduce the beam strength of the pipe. Because there is already a stress on the pipe this can lead directly to premature failure of the pipe by fracturing.

The fracture of a cast iron main will usually result in a sudden release of gas which can, where site circumstances are such, lead to gas ingress into adjacent property and potentially result in an explosion. It was the increase in such events that led to the introduction of a number of replacement programmes that are now described.

2.1 Early mains replacement- pre 1977

BG Plc and the former British Gas Corporation (BGC) have had a mains replacement policy since soon after the company was created in 1972. At this time BGC was made up from twelve previously autonomous Regional Gas Boards. Originally each mains or towns network belonged to small individual local gas companies. These networks began to be integrated soon after nationalisation of the gas industry in 1957.

Prior to 1974 there had been no national policy for mains replacement and generally, mains were replaced because of their poor condition or in association with reinforcement. About 600 km within the system had been replaced each year in the preceding decade. There was concern at the level of incidents occurring and in 1974 an interim national policy was introduced for the replacement of mains and services to reduce the risk of incidents. Mains replacement was set at approximately 1% of the total population per annum, (1,920km), for the next 10 years based on a qualitative assessment of the main.

2.2 King Replacement 1977-1984

Several severe gas explosions occurred over the Christmas & New Year period of 1976/7. This was due in part to the particularly cold weather conditions and buildings being unoccupied over the Christmas/New Year holiday period, allowing gas to build up undetected. As a result, the Secretary of State for Energy commissioned an inquiry chaired by Dr. P. J. King, to examine the circumstances surrounding the incidents and to consider improvements to existing procedures and systems, which might reasonably be implemented and lead to a reduction in such incidents.

Although the laying of cast iron mains had ceased some years before the inquiry, in the late 1960's, at that time about 80% of the distribution system consisted of this material. Rapid replacement of the whole system was not a practical option and it was recommended that the replacement of "higher risk priority mains" should be carried out by the end of 1984.

The mains targeted were small diameter cast iron and steel mains in the most hazardous locations. Priority was given to those mains operating at higher pressures where significant quantities of gas would be released as a result of a fracture. The definition of hazard locations was based around topographical features, including the amount of unsurfaced ground between the main and building, the presence of cellars, proximity and type of adjacent buildings where the likelihood of gas ingress to properties was greater. This resulted in an increase in the rate of replacement to a level of 2,800 kms per year.

2.3 Post King Replacement 1985-1990

The need to continue a programme of mains replacements to contain or reduce hazard, was recognised in the King Report and confirmed in the Corporation's policy objective that the replacement activity should be "maintained after 1984 until all secondary risk mains have been eliminated".

In 1980 a study was made to identify these secondary risk mains for replacement. An analysis was made of serious incidents and mains breakage rates. Based on this analysis, the expected level of incidents and fractures at the end of the King Programme was predicted. It was concluded that to maintain forecast safety levels, it would be necessary to abandon mains at a rate of 2,600 km/annum until the end of 1989. The analysis of mains system deterioration through broken mains and incidents indicated that the greatest safety "benefit" would arise from abandoning mains in certain broad categories. This led to the continued replacement of cast iron mains in the smaller diameter ranges but extended to include mains up to 7" diameter. Priority was again given to those mains in the most hazardous locations and those operating at higher pressures.

A further review in 1985 (Marchant) resulted in the criteria used to determine replacement being expanded. It had been based primarily on mains diameter, material and hazard location. The new criteria placed increased emphasis on breakage/corrosion history, evidence of subsidence, soil conditions and traffic loading. Some of these features were included in an early points scheme which had been developed and was being used in some of the regions.

Up to this time, the replacement programme was generally reactive, based on simple analysis of the history of incidents and latterly fractures and gave rise to the replacement of mains in broad categories. In addition, mains continued to be replaced because of their condition, where the cost of replacement could be balanced against the continued cost of repairs or where the main was no longer fit for purpose.

2.4 Points Scheme 1990-present

As mentioned previously, during the early 1980's BG began to develop a scheme based upon assessing each mains unit separately, rather than considering groups of mains. This was for the cast iron mains population with diameters less than or equal to 12". At this time, the parameters considered to affect the likelihood of failure leading to an explosion incident included the diameter of the main, breakage of surrounding mains, previous fractures on a main, the distance between the main and nearby buildings, the existence of cellars in nearby buildings, and the operating pressure of the main.

The basis of the scheme, known as the Points Scheme, was to model the process leading to an explosion incident, namely mains fracture, gas ingress into a building, and ignition of the accumulated gas. At each stage, different characteristics of the main were considered to contribute to the likelihood of that event occurring. Each characteristic of the main (diameter, pressure etc.) and its environment (proximity, existence of nearby cellars etc.) attracted a given value. A final 'score' was generated by combining all of these values together for each mains unit.

In order to produce a prioritised list of mains units for replacement, each individual unit had to be assessed. This involved collecting information on all one million cast iron units, by carrying out on-site assessments. This survey process was carried out during 1988 and 1989, in order that the new scheme was ready to implement in January 1990. Once all survey data was collected, analysis was carried out to determine a replacement threshold, above which mains were earmarked for replacement. This threshold was set at 1200 points in order that incident levels should fall from an average of 6 per year in 1990 to an average of 3 per year by 1995.

Thus the replacement policy in operation from 1990 onwards was to replace all mains units with a score exceeding 1200 points, by December 1995, although it was recognised that some regions would go beyond this due to the large numbers of mains above the threshold, and an extension to 1998 was given. Throughout the period, some mains would fracture, thereby increasing their points score. If a particular main moved from below 1200 points to above, with the addition of a fracture, it would then become a candidate for replacement. This process is known as dynamic growth, and will continue, even when all mains above a given threshold have been replaced.

The replacement of mains above the 1200 points threshold was not quite achieved due to the dynamic growth element of the system. It was recognised that significant changes may have occurred in the 'on site' conditions that existed when the mains were first surveyed some years previous. As a result, a programme of re-survey of the mains system was undertaken to capture these changes and reassess the points score of the mains unit. This has resulted in a significant replacement workload.

It was recognised that whilst the Points Scheme had served the company well in its selection of mains for replacement, there were elements within the calculation that were subjective and there was some duplication of effect within the factors used to derive the score. The numbers of ductile iron failures due to corrosion were also increasing.

It was decided to conduct a review of the Points Scheme to make it more risk based and at the same time to consider a scheme to assess the risk presented by the ductile iron population and the cast iron mains with diameters greater than 12". The proposal was discussed with the safety regulator, the Health and Safety Executive (HSE), who were keen to be involved in the review from the outset. It was essential that the new models should be capable of at least sustaining the existing levels of safety

3. Development of new risk assessment scheme

3.1 Requirements of new scheme

At this time the company was entering discussions with Ofgas (now OfGem) on its Price Review Formula. In order to satisfy the requirements of both Ofgas and the HSE, Transco were required to demonstrate a quantitative level of safety from the distribution system, arising from a specified level of replacement expenditure. The Points Scheme is essentially a prioritisation scheme, i.e. it allows mains to be identified in order of decreasing risk. However, because of the limited amount of historical data upon which the scheme was based, it could not be demonstrated that a main of 1000 points was necessarily twice the risk of a main with 500 points for example, only that it was of higher risk. Thus an assessment scheme which could quantify risk on an individual unit, and hence from the system as whole, had to be developed. The development of this new scheme was undertaken by BG Technology on behalf of Transco.

3.2 Basic structure

The basic structure of the new scheme is the same as that of the Points Scheme, namely the product of a three stage process of mains fracture, gas ingress, and ignition. During the period in which the Points Scheme has been operating, Transco has been collecting much more information on fractured mains, in particular the characteristics of a fractured main and its environment, together with the consequences of the fracture, i.e. whether the leak tracked into nearby buildings.

From analysis of this large quantity of historical data, it was possible to quantify the effect of particular characteristics of the main and its surroundings much more accurately. BG Technology were then able to use these results to build up a model, which ultimately calculates a risk value, in terms of explosion incident per km per year, for each individual mains unit.

3.3 Factors to be included

Most of the parameters which had been included in the Points Scheme were retained within the new scheme, with the addition of one new parameter, namely the amount of open (unsealed) ground between a main and the nearest property. From analysis of historical data covering the previous seven years, there appeared to be a strong relationship between the amount of open ground and the likelihood of gas ingress arising from a fracture.

The presence of a grass verge between a main and nearby property will tend to reduce the likelihood of gas tracking underground from a fracture, because the gas has a route to atmosphere, whereas a sealed surface, such as a concrete drive would tend to increase the likelihood of tracking. Detailed information from previous fractures allowed this effect to be quantified accurately. In addition, the contribution of other existing parameters was also refined greatly.

3.4 Background Breakage Zone

The Points Scheme had included a parameter called the Background Breakage Zone (BBZ). This took into account fractures occurring on other mains in the vicinity of an individual main, and was used to take account of the effect of ground subsidence etc, in the neighbouring area. If a main had not fractured, but the fracture rate of nearby mains was high, the risk value of the individual unit was elevated to take account of the fact that local ground conditions would potentially have an effect on that particular main in the future.

Each region of Transco calculated BBZ's using available information. This was usually a time-consuming manual process and partly subjective. With the advent of a national register of digital records, containing precise information about the location of mains, BG Technology has developed a BBZ model which performs sophisticated cluster analysis which combines the geographical data and fracture records, to provide a consistent and accurate measure of BBZ for the whole cast iron system. Because this is now automated, it is intended to update BBZ's on a regular basis to take account of changes in the mains population due to re-breakage.

4.0 Further assessment models

Approximately half of Transco's existing distribution system is made up of Cast Iron mains. The risk scheme described in section 3 of this paper is applied to Cast Iron mains with diameters up to and including 12 inches. Mains above this diameter have a greatly reduced rate of fracture, and thus the amount of historical data is much more sparse. In addition, the mode of failure for large diameter Cast Iron may be different from that for smaller diameters. For these reasons, a separate model was developed by BG Technology to assess these large mains. It uses some elements of the smaller diameter scheme together with theoretical modelling techniques.

In addition, Ductile Iron mains, which make up about 10% of the distribution system, have also been considered. Ductile Iron fails through corrosion, and the release of gas from the resulting corrosion hole can be equivalent to that released from a fractured cast iron main. In addition, a greater proportion of Ductile Iron mains operate at higher pressures than Cast Iron mains. A separate scheme for Ductile Iron uses part of the methodology from the Cast Iron Scheme, together with corrosion data, to predict incident rates arising from corrosion of Ductile Iron mains. Once again, this approach relies on historical data collected by Transco on Ductile Iron failures and their consequences.

All three models generate output in terms of explosion incidents per km per year for each mains unit, and are thus compatible with one another. This is an important factor when considering the implementation of such schemes, as Transco needs to be able to manage mains replacement without considering each material separately.

5.0 Implementation

Once all mains have been assessed using the new schemes, Transco will have a national list of mains, and will be able to manage its replacement on the basis of quantified risk. Work is currently being carried out to determine a suitable threshold for replacement. This threshold will be set to ensure that the level of incidents likely to arise from the remaining mains population does not increase above an agreed safety level.

Once the threshold has been calculated, the length of main identified for replacement and hence the expenditure required, can be determined. Thus, the output from the models can be used to set replacement policy for the future.

Transco are currently in the process of collecting survey information required as input to the new models. In the meantime, survey data collected for the Points Scheme can be converted for use in the new small diameter cast iron scheme. A diagram representing a typical street layout, with a mains unit and surrounding buildings is contained in Figure 2. Once all data has been collected, Transco will operate a national scheme, identifying mains for replacement above the stipulated threshold.

Transco consists of twelve Local Distribution Zones (LDZ's). Information will be available for each LDZ, in order that replacement expenditure can be managed locally. Each LDZ will have a different allocation of mains to replace, depending on how much main currently lies above the set threshold. In addition, each LDZ will have a certain amount of dynamic growth to consider, based on the quantity of mains fractures within a given LDZ, and changes in local environment, such as the change of open ground to sealed ground as householders convert grassed areas to hard standing.

6.0 Conclusions

It can be clearly demonstrated that the policies adopted have been very effective in reducing the numbers of mains related incidents even though they were developed from the limited data that was available at that time (see Figure 3). This has been done by targeting mains in broad categories which presented the greatest risk.

From the early 1980's additional data was collected each time a main fractured. This data has been used to develop the models used to enhance the targeting of mains for replacement in subsequent policies. The Points Scheme has enabled an assessment to be made of all the small diameter cast iron mains units allowing the engineer to prioritise their replacement. The new risk based models will allow all the cast iron and ductile iron mains of all diameters to be assessed and their replacement programmed according to the risk they present.

In order to understand the factors that influence the failure of mains and the potential to cause incidents, it will be necessary to continually review the models and update them as more data becomes available. In this way the risks posed by the mains network can be minimised.

Clearly, despite their age, a number of cast iron mains will be fit for purpose for some time to come and present little or no risk. Others will need to be replaced on a condition basis to ensure the integrity and continued safe operation of the distribution network. Development of a model to assess the replacement of mains on a condition basis is currently being undertaken.

Using the outputs of all these models, it will be possible to develop a replacement strategy which targets the right mains for replacement at the right time and thereby will satisfy the requirements of both the safety and financial regulators.

FIGURE 1

TABLE OF MAINS POPULATION BY AGE (km)

Decade	Pit Cast/ Sand Spun	Metal Spun	Ductile Iron	Steel	PE	TOTAL
Pre 1880	193	0	0	0	0	193
1880-1889	320	0	0	0	0	320
890 - 1899	882	0	0	0	0	882
900 - 1909	4753	0	0	102	0	4855
910 - 1919	2831	0	0	79	0	2910
920 - 1929	12219	668	0	49	0	12936
930 - 1939	18703	5773	0	148	0	24624
940 - 1949	2298	7686	0	569	0	10553
950 - 1959	807	29906	103	2019	0	32835
960 - 1969	229	23157	5031	5028	101	33546
970 - 1979	0	1261	14465	3790	19027	38543
980 - 1989	0	18	1993	1917	52488	56416
1990-1995	0	0	18	198	30384	30600
TOTAL	**43235**	**68469**	**21610**	**13899**	**102000**	**249213**

FIGURE 2

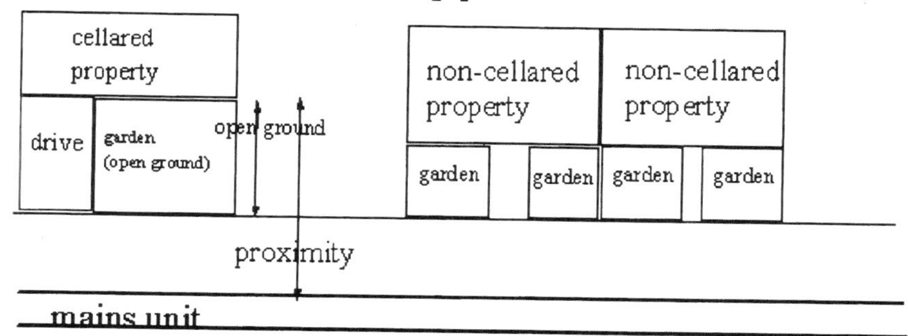

FIGURE 3

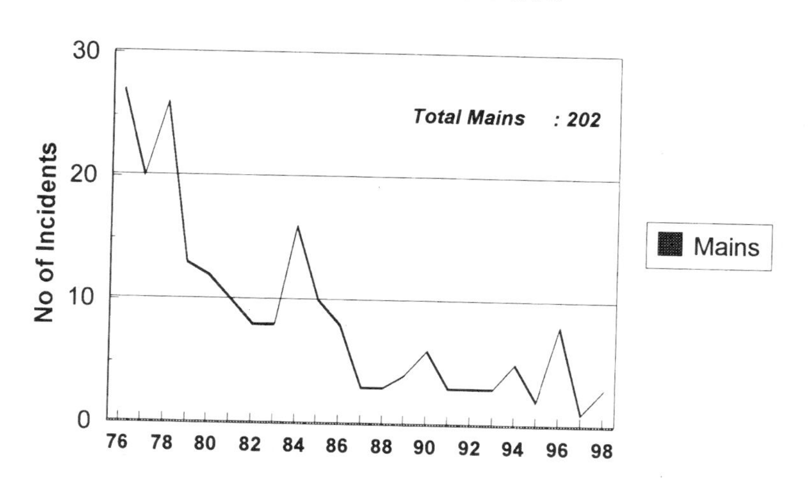

C571/019/99

Decision-making analysis for renovation of potable water mains based on pipe sample data

A GOUGH
Pipelines Department, Charles Haswell & Partners Limited, Birmingham, UK

ABSTRACT

Charles Haswell and Partners Limited is currently undertaking a large programme of water mains rehabilitation design for Severn Trent Water Limited. Around 1989 Severn Trent embarked upon a major mains renovation programme; some 80% of the mains at that time were ferrous materials. Corrosion was, and still is, the main enemy of the distribution system. The corrosion process impacts in many ways causing structural instability and internal restrictions (tuberculation). This paper describes how Severn Trent uses physical pipe samples to determine the effects of corrosion on the ferrous pipes. The information collected from the samples is used to evaluate the most cost effective rehabilitation method and for asset planning. The paper will also outline proposals to use the Company's extensive historical data as a predictive tool in lieu of taking physical samples.

1 INTRODUCTION

Severn Trent Water Limited provides water and sewerage services to a population of eight million people.

The Company has a mains distribution network of c. 40,000 kms in length. The majority of the mains are manufactured from ferrous materials, most commonly cast iron. Figure 1. A major mains rehabilitation programme has been underway since 1989. The triggers for this work are water quality, appearance, taste, structural condition, leakage and maintenance costs.

The Rehabilitation Engineer has two major decisions to make:

- when does a water main require rehabilitation?
- what rehabilitation method would be most cost effective?

It is not the intention of this paper to answer the former in any great detail; however, it will describe how pipe sample data can assist with the latter.

The simplistic choice of rehabilitation methods is between relining and renewal. Structurally sound ferrous mains are internally lined in situ with cement mortar or epoxy resin. These linings form a protective barrier between the water and the pipe wall to arrest the internal corrosion process. However, the linings are non-structural so they rely upon the inherent strength of the existing main to continue in service. Furthermore, the lining process does not reduce, or prevent, leakage. The cost of lining existing mains can be significantly cheaper than that of total replacement. However, lining a structurally unsound main, or one with high leakage, may be uneconomic when considering whole life costs.

Unsound mains should therefore be considered for renewal.

The process that is predominant in rendering ferrous mains unsound is corrosion.

2 CORROSION

The effects of corrosion on potable water mains usually manifest themselves as distribution problems such as leakage, water quality and appearance; the very drivers of the rehabilitation programme. As far as the Rehabilitation Engineer is concerned corrosion has two main consequences; the migration of iron from the structure leaves a weakened graphitic shell; the products of corrosion congeal to form internal protrusions (tuberculation) which affect hydraulic performance. Corrosion also has economic effects. Figure 2.

Whilst corrosion undoubtedly affects cast iron, the predominant material, the process is complex and is dependent on many factors. Therefore, the existence of a cast iron main in itself is no reason to rehabilitate.

3 USE OF CAST IRON

The earliest use of grey cast iron, as a material, was to cast a lion in China some 2,500 years ago; the casting still remains relatively unaffected today. The first recorded use of cast iron pipe to transport water was in 1455 to supply water to Dillenburg Castle, Germany. Some fifteen miles of cast iron pipe laid in 1664 to supply the fountains in Versailles, France are still in use today (1).

These examples eloquently demonstrate that not all cast iron mains require rehabilitation, or replacement, unless there are some indicators to the contrary.

A practical option is to physically inspect and test ferrous mains to determine if there is a need to rehabilitate.

4 USE OF PIPE SAMPLES IN SEVERN TRENT WATER LIMITED

4.1 Pipe Samples

Historically, evidence to support the structural condition of water mains typically came from burst main records, anecdotal reports, and targeted or opportunistic visual inspections. The disadvantage of these methods is that they can be subjective, inaccurate and unsubstantiated.

Figure 12 shows a typical result from the programme. Both types of predicted life expectancy, Staffordshire University (SUL) and WRc (WRL), are given together with the life after relining (RAL).

The development of the programme is focusing on obtaining better correlation by refining some of the variables. For instance, water type is being analysed for pH, alkalinity and conductivity when considering internal corrosion. Further development will study ground conditions and burst main history in relation to external corrosion and mains failures.

6 CONCLUSIONS

Severn Trent Water has demonstrated over many years experience that physical inspection and analysis of pipe samples is a positive aide to making rehabilitation decisions. The data can be used to determine local or wider asset management solutions. Pipe sample data is particularly useful when considering mains renewal solutions over non-structural lining. The extensive data already gathered, and continuing to be collected, could be used for predictive purposes eliminating the need to collect expensive pipe samples. Further development of the model is required to improve the correlation between corrosion factors and the rate of corrosion and tuberculation growth.

REFERENCES

1. G.Gedge, Corrosion of Cast Iron in Potable Water Service, The Society of Chemical Industry, London, December 1992

2. A.M.Dean, H.C.Lowe and A.P.Parker, An Investigation into the Corrosion, Strength Reduction and Flow Resistance of Cast Iron Water Mains in the Potteries, Moorlands and Stafford Areas of Severn Trent Water Authority, Stafford 1984, North Staffordshire Polytechnic.

3. G.Bancroft, Severn Trent Project, Statistical Analysis of Existing Data, School of Computing, Staffordshire University.

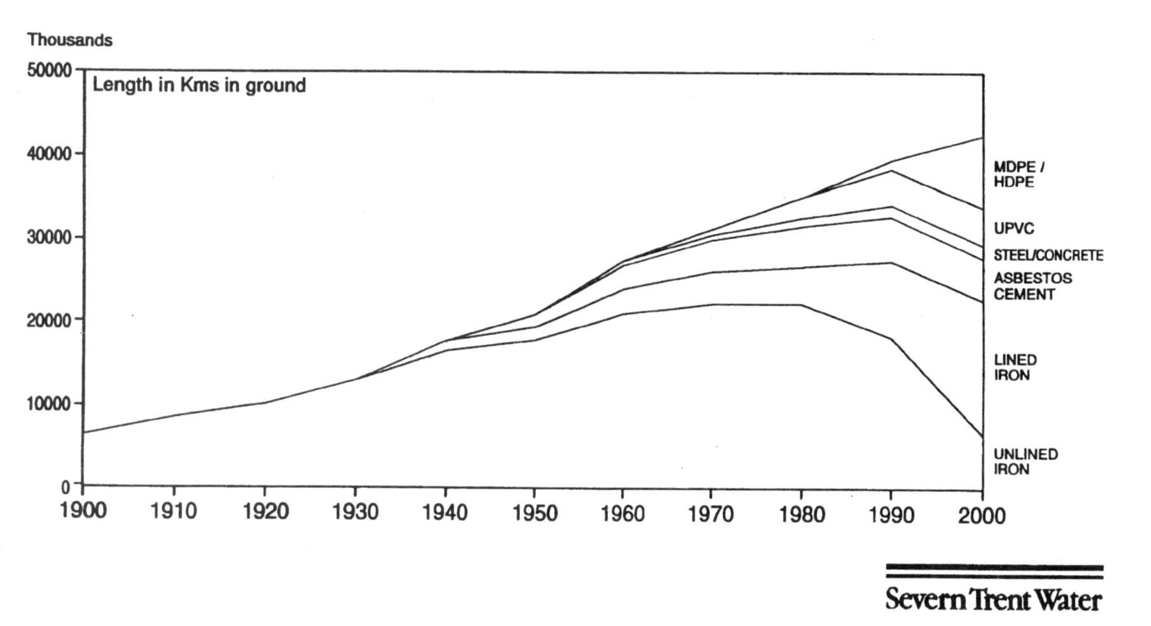

Figure 1. Cumulative length of water main

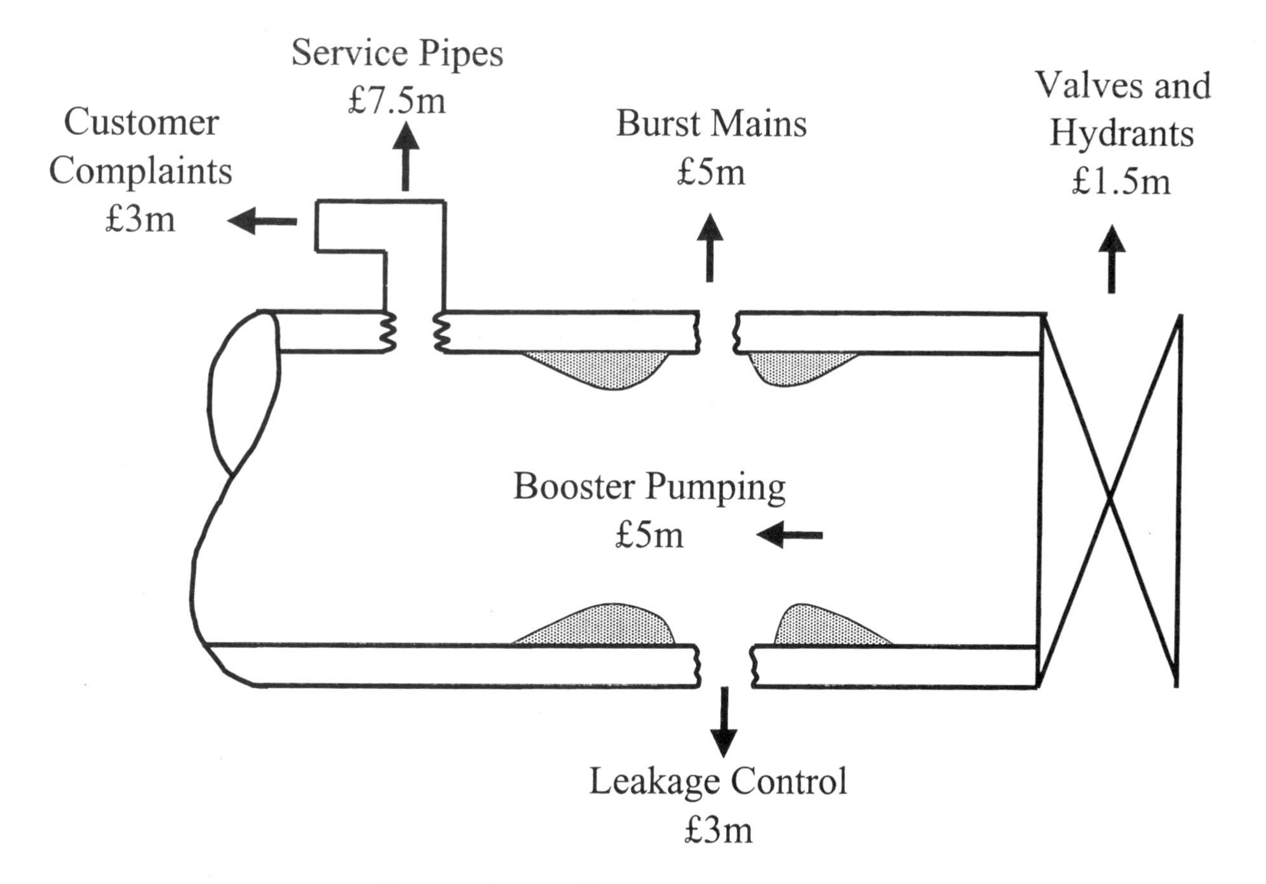

Figure 2. Indicative costs of corrosion

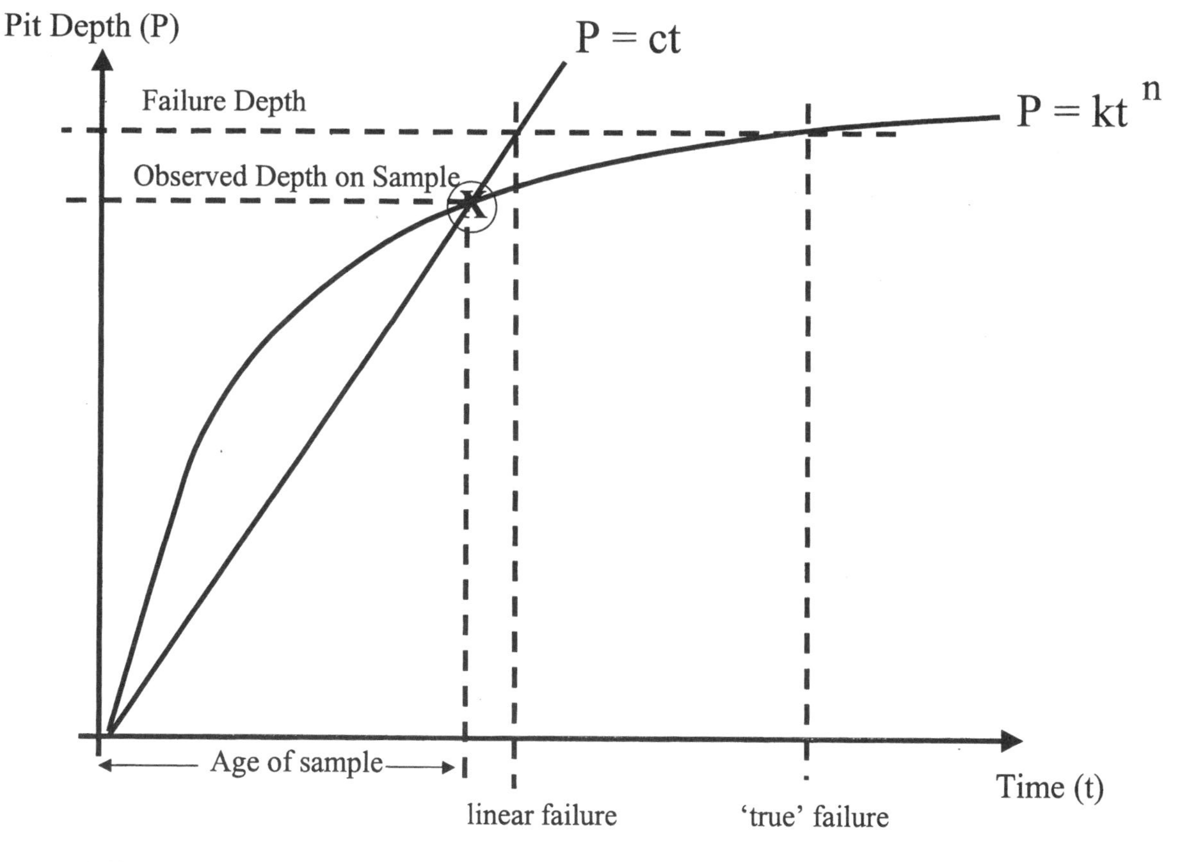

Figure 3. Assumed v actual corrosion rates for spun iron

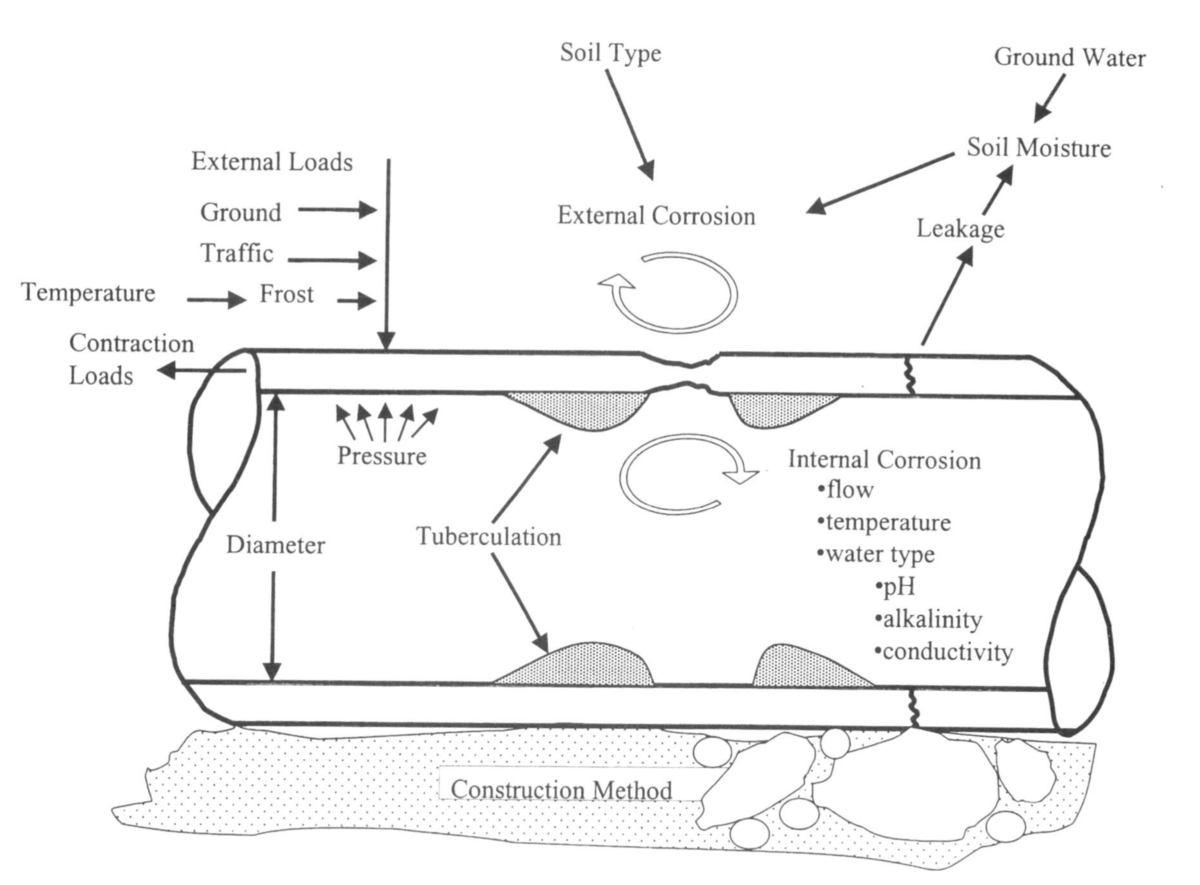

Figure 4. Factors affecting corrosion and structural condition

Staffordshire
UNIVERSITY

STW SERIAL NO
0870112

SCHOOL OF ENGINEERING

SEVERN TRENT LTD
REGIONAL SURVEY OF CAST IRON WATER MAINS

TEST REPORT SHEET

STW SERIAL NUMBER									
0	8	7	0	1	1	2			
MANAGE-MENT DISTRICT		DMA			SAMPLE NO		SUFFIX		

BATCH/REPORT NO			
S	Ø	2	4

1. **LOCATION DETAILS**

GREYS GARAGE, WHARF ST,
WARWICK.

2. (a) **PIPE CLASSIFICATION**

Year Laid: 1934

Nominal Pipe Diameter (Tick Box)

	2 in	3 in	4 in	6 in	8 in	>8 in
		✓				
OC	2	3	4	6	8	998

Figure 5. Extract from pipe sample report

STW SERIAL NO
0870112

9. CORROSION ASSESSMENT

DAMAGE	EXPOSED SURFACE	SECTION THROUGH EXPOSED SURFACE	EXTERNAL	INTERNAL	OC
Unspecified					0
Even General					1
Uneven General					2
Even Local					3
Uneven Local			✓	✓	4
Pitting					5
Cracking					6

10. PIT DEPTH/WIDTH MEASUREMENTS

External (to 1 dec place)

CROWN POSITION ~~MARKED~~/ESTIMATED (delete as appropriate)			INVERT		
	PIT DEPTH (mm)	PIT WIDTH (mm)		PIT DEPTH (mm)	PIT WIDTH (mm)
1	2·3	13	1	1·8	2
2	3·5	3	2	0·8	2
3	2·2	12	3	1·2	8
4	1·9	10	4	0·6	2
5	2·5	8	5	1·5	6

Figure 6. Extract from pipe sample report

15. TUBERCULATION MEASUREMENT

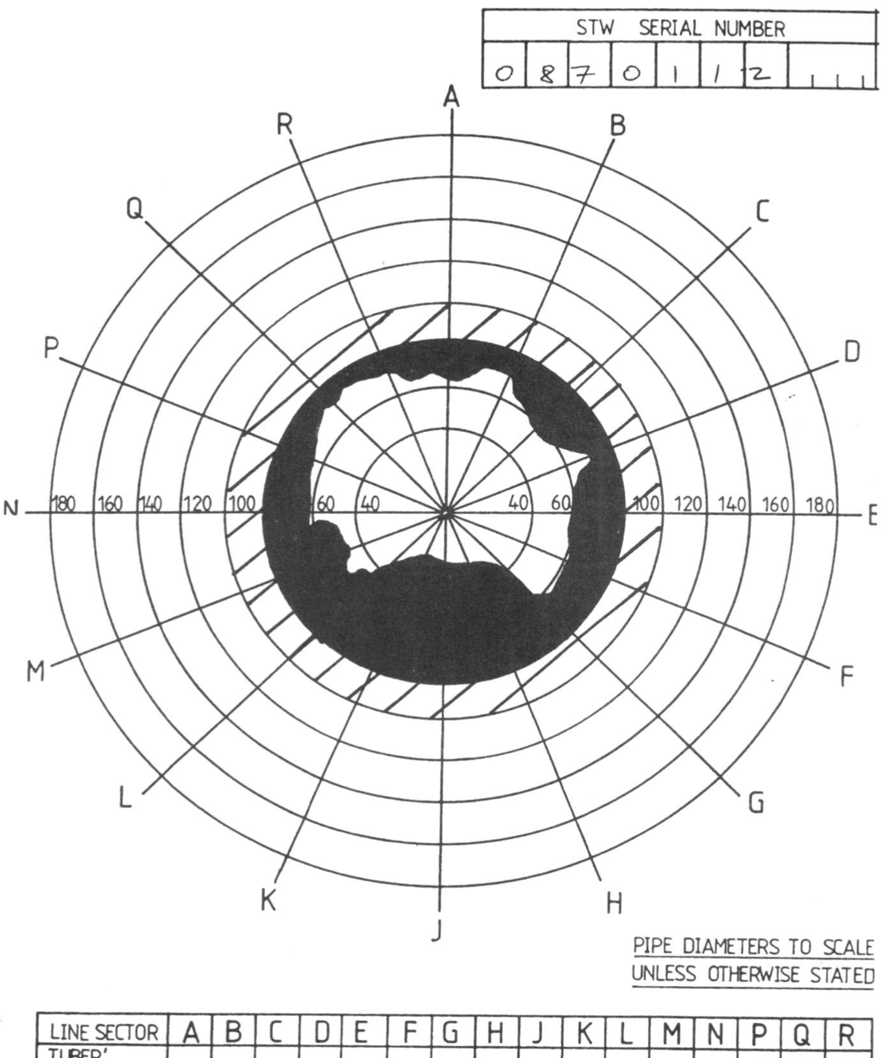

LINE SECTOR	A	B	C	D	E	F	G	H	J	K	L	M	N	P	Q	R
TUBER' HEIGHT (mm)	10	7	13	6	13	12	20	29	30	31	25	19	12	10	6	6

Figure 7. Extract from pipe sample report

TABLE 1 - SUMMARY OF RESULTS   10-MAR-95   PAGE: 1

Where appropriate dimensions in mm
Life value of -1 indicates infinite life

Priority		-1 Perforated on receipt		0 Failed during testing		-99 Indeterminable	
Gradings (PRIO)		1 < 5	2 5-10	3 10-20	4 20-30	5 30-40	6 40-50
		7 50-75	8 75-100	9 100-200	10 >200	99 Indefinite life	
Material (MAT)		1 Vertically Cast	2 Spun Grey Iron	3 Ductile Iron	4 Other		
Water Type	0 Unspecified	1 Upland Res	2 Upland River	3 Lowland River	4 Borehole	5 Other Bulk Supply	
Soil Type	0 Unspecified	1 Clay	2 Sand/Gravel	3 Rock	4 Slag/Ash	5 Soil	6 Pottery Waste

ALL DATA, INCLUDING AGE AND LIFE EXPECTANCIES, AS ON DATE OF SAMPLING

Where_batch=S024

STW CODE	PRIO	MEASR INTNI DIAM (mm)	AVERG WALL THICK (mm)	MAT	AGE (yrs)	DATE SAMPLED	EXT PIT DEPTH (mm)	EXT CORR RATE (mm/yr)	INT PIT DEPTH (mm)	INT CORR RATE (mm/yr)	THIN WALL REMAI (mm)	LIFE SP (0 yrs	LIFE WRC mm yrs	LIFE AFTR RELN FAIL) yrs	AVE TUB HGHT (mm)	TUB PER YEAR (mm/yr)	PERC BLKD	SOIL	WATER	REPORT NO
0810003	7	102.5	9.0	2	45	05-JAN-95	3.4	0.076	2.8	0.062	5.6	41	74	74	0.0	0.000	0.0	0	3	S024
0811501	5	80.5	7.7	2	40	05-JAN-95	1.7	0.043	3.4	0.085	2.6	20	31	61	20.8	0.520	76.6	1	3	S024
0820806	4	73.5	10.5	1	55	05-JAN-95	2.1	0.038	5.7	0.104	3.0	21	29	79	10.2	0.185	47.8	5	3	S024
0822313	9	235.0	11.9	2	35	05-JAN-95	0.0	0.000	2.9	0.083	8.6	104	104	-1	2.3	0.066	3.9	2	3	S024
0830110	9	104.0	9.8	2	40	05-JAN-95	0.0	0.000	2.4	0.060	10.1	168	168	-1	1.4	0.035	5.3	1	3	S024
0830111	8	79.0	7.7	2	30	05-JAN-95	1.9	0.063	0.6	0.020	5.5	66	87	87	0.3	0.010	1.5	2	3	S024
0832311	7	106.0	7.7	2	33	05-JAN-95	1.2	0.036	2.6	0.079	5.1	44	65	140	3.3	0.100	12.1	1	3	S024
0870112	5	82.5	8.1	1	61	05-JAN-95	3.5	0.057	5.1	0.084	2.8	20	33	49	15.6	0.256	61.3	2	3	S024
0885201	5	83.0	7.2	2	37	05-JAN-95	1.2	0.032	3.5	0.095	3.3	26	35	102	24.9	0.673	84.0	3	3	S024

9 rows selected.

Figure 8. Pipe sample results

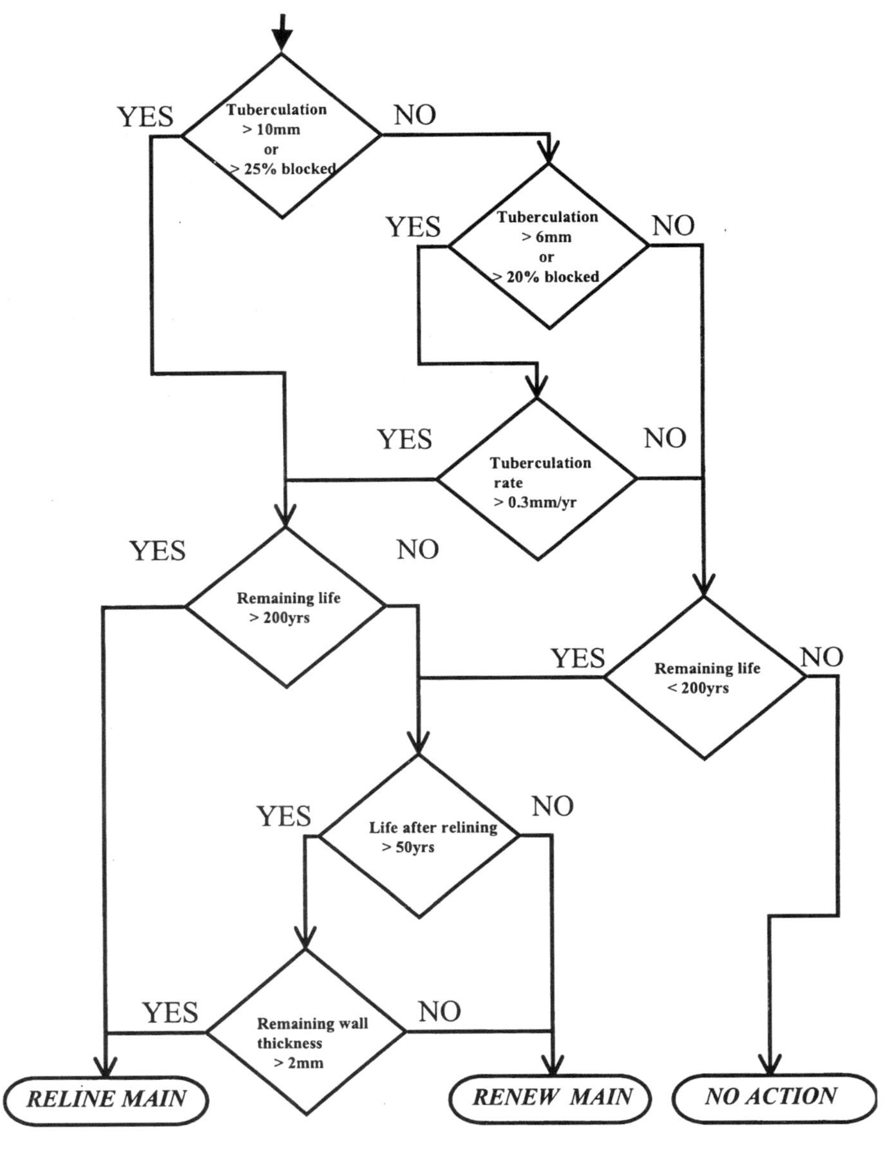

Figure 9. Pipe sample decision diagram

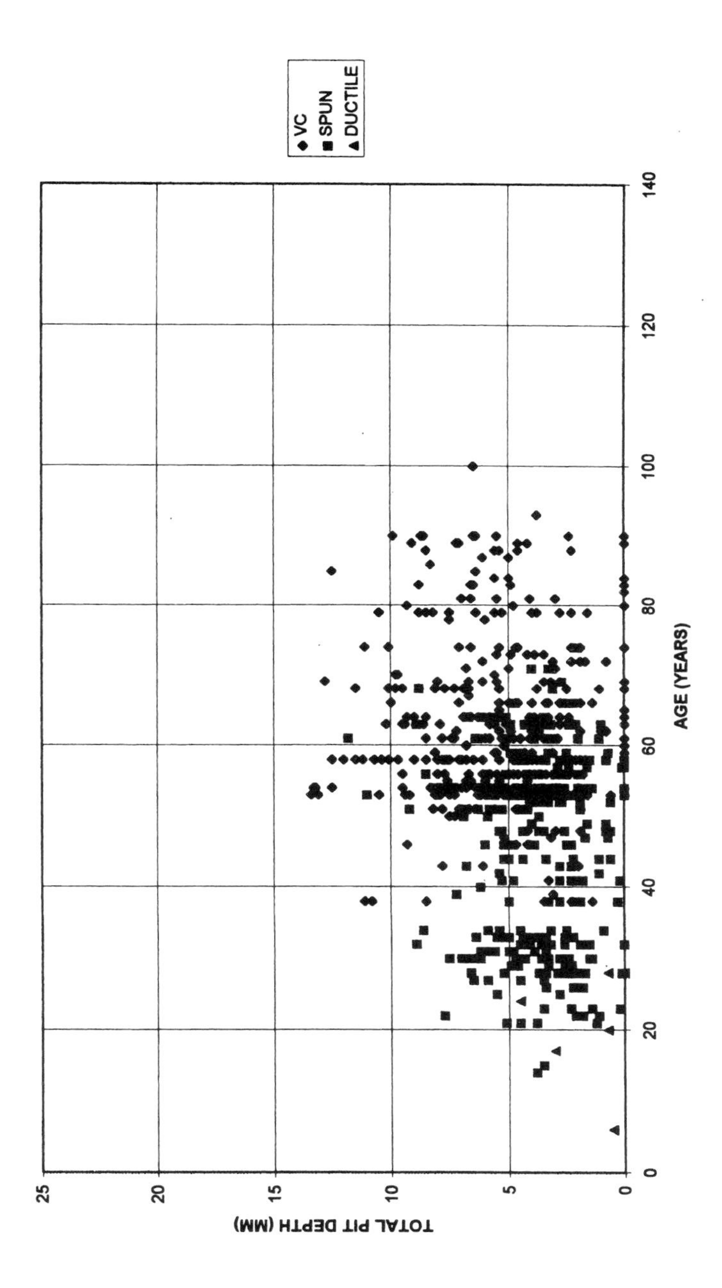

Figure 10. Total pit depth versus age for Nottingham

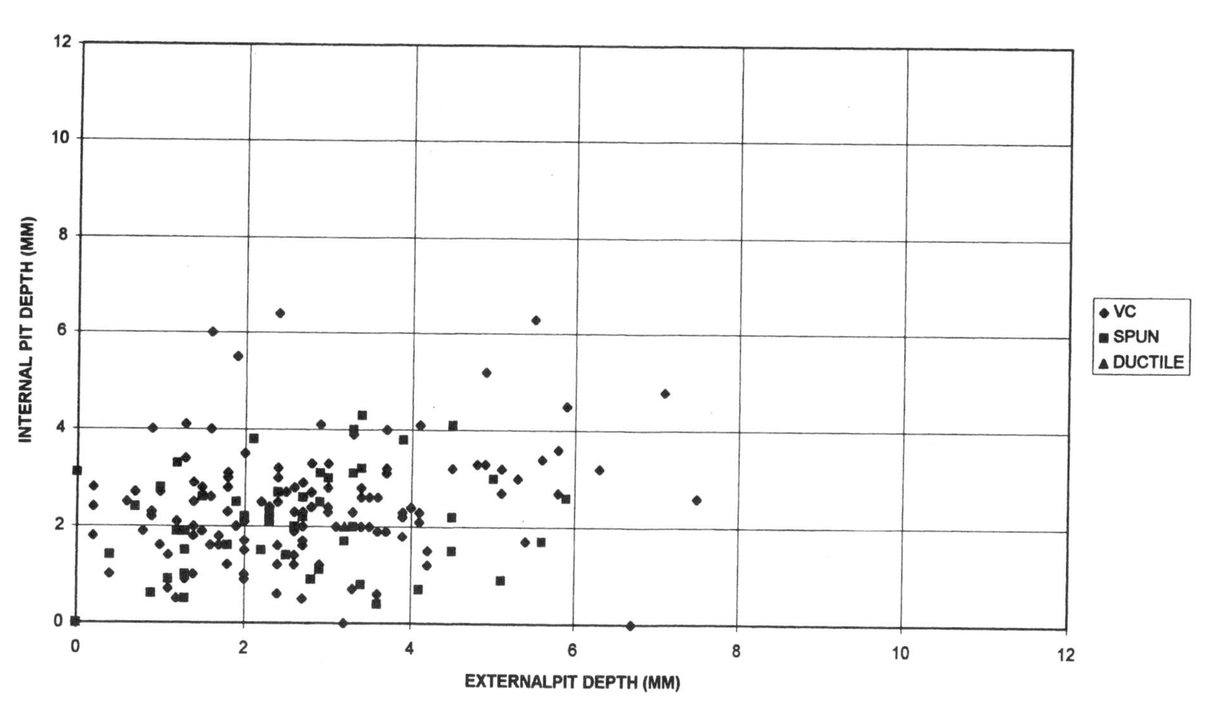

Figure 11. External versus internal pit depths for Wolverhampton

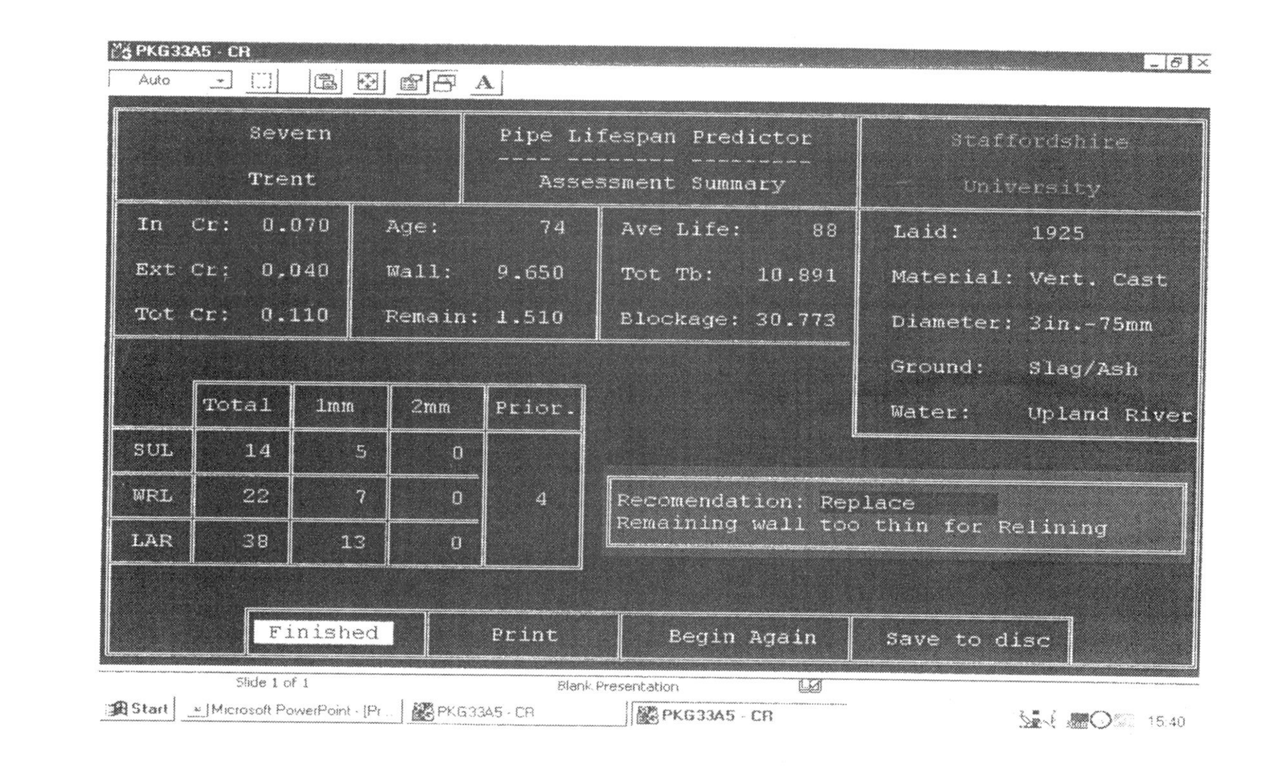

Figure 12. Typical result from predictor programme

Low Pressure: Safety and Integrity

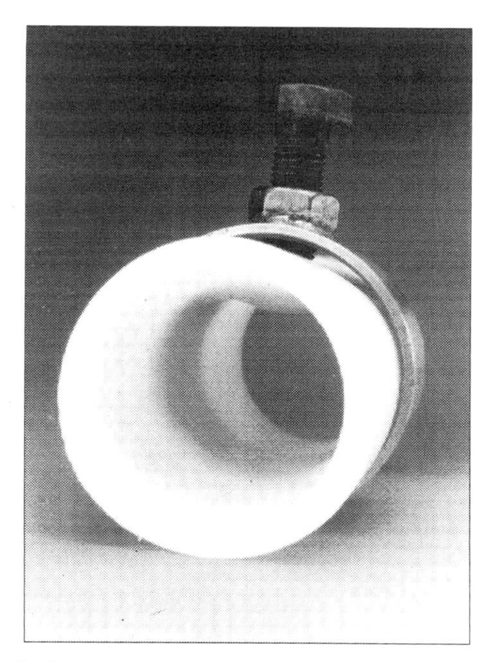

Figure 1: Reversion of Cross-Linked Polyethylene around obstacle

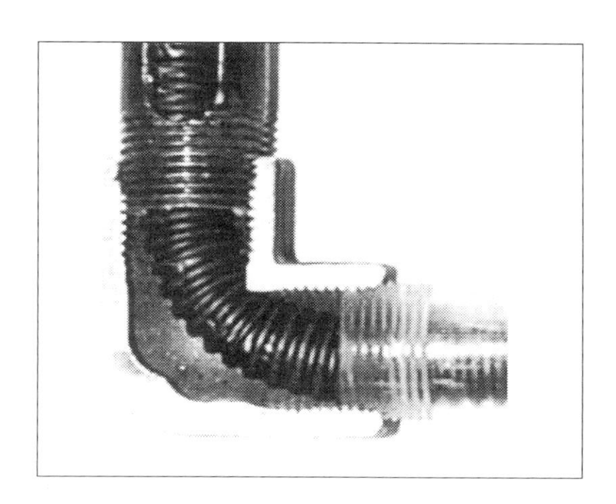

Figure 2: Serviflex pipe in knuckle elbow

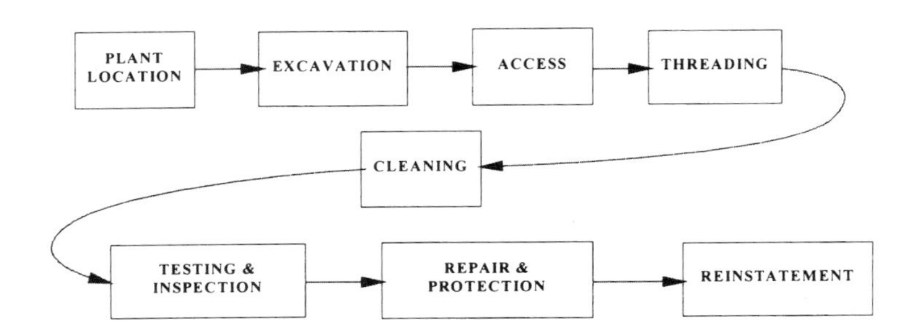

Figure 3: Pipeline Maintenance Process Model

Figure 4 – Under Pressure Drilling

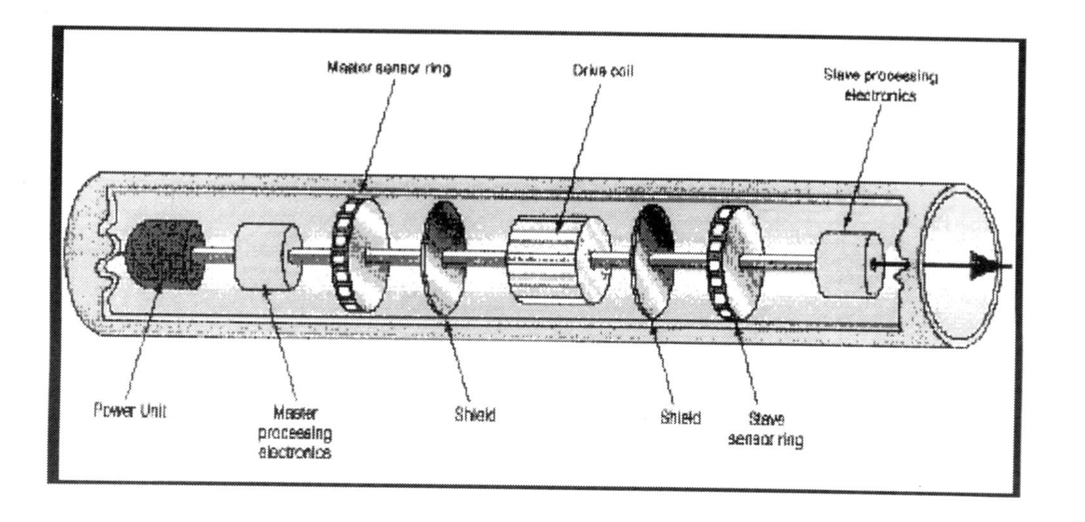

Figure 5: Remote Field Eddy Current Tool Schematic

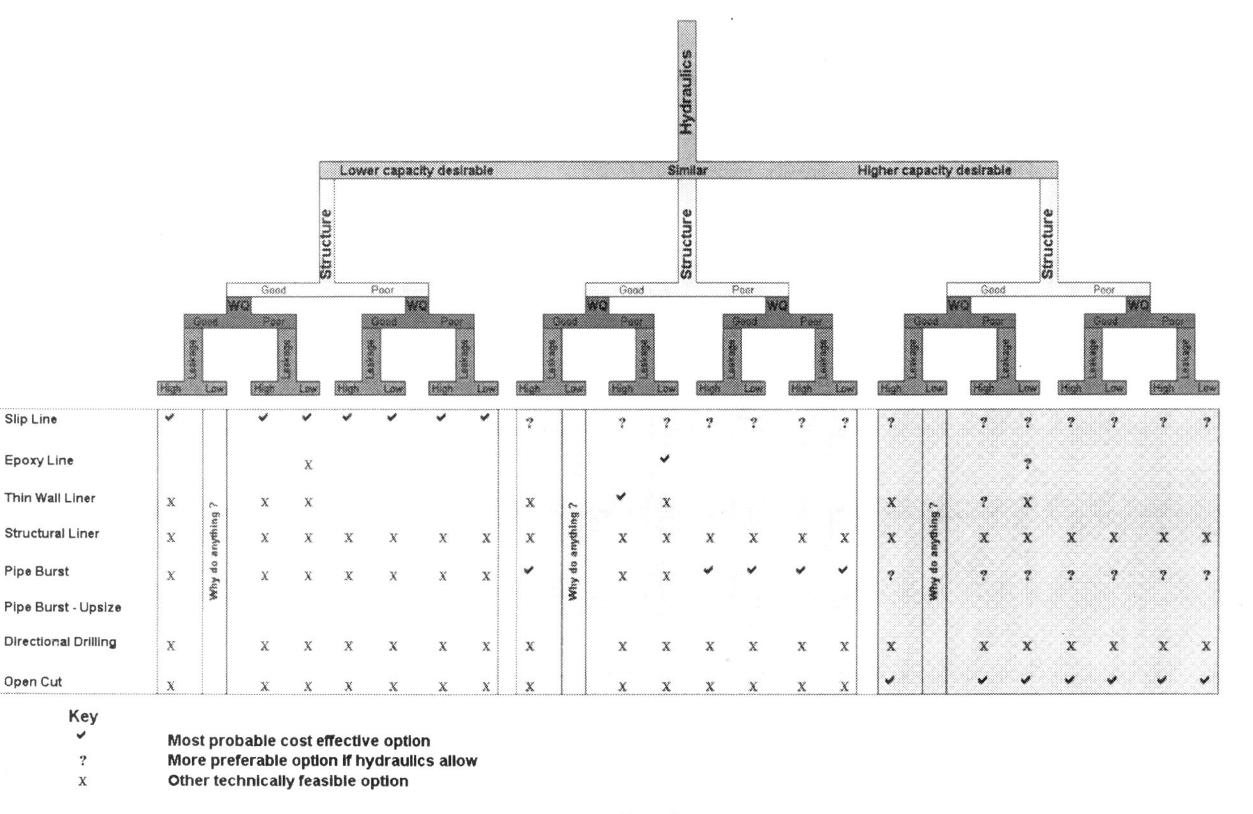
Hydraulics
Lower capacity desirable
Similar
Higher capacity desirable
Structure
Good
Poor
WQ
Leakage
High
Low
Why do anything ?
Slip Line
Epoxy Line
Thin Wall Liner
Structural Liner
Pipe Burst
Pipe Burst - Upsize
Directional Drilling
Open Cut
Key
✓ Most probable cost effective option
? More preferable option if hydraulics allow
X Other technically feasible option

Figure 8